And the Rest is Just Algebra

Praise for "*And the Rest Is Just Algebra*"

Middle school, college algebra, calculus, and linear algebra teachers can all benefit from this book. Becoming proficient with algebra is a complex task—much more so than it appears to those who were successful the first time around. Furthermore, according to neuroscience studies, manipulating symbolic expressions requires considerable cognitive effort even for those who are proficient. In addition, "met-befores" in the form of prior arithmetic thinking can thwart algebraic thinking. All this, and more, is considered in a readable way in this book.

Annie Selden

Sepideh Stewart
Editor

And the Rest is Just Algebra

Springer

Editor
Sepideh Stewart
Department of Mathematics
University of Oklahoma
Norman, OK
USA

ISBN 978-3-319-45052-0 ISBN 978-3-319-45053-7 (eBook)
DOI 10.1007/978-3-319-45053-7

Library of Congress Control Number: 2016953106

Printed on acid-free paper

This Springer imprint is published by Springer Nature
The registered company is Springer International Publishing AG Switzerland

Foreword

The mathematics education community has had a keen interest in students' learning of algebra, especially since the "Algebra for All" movement began in the 1990s. The great bulk of research on students' algebra, however, focuses on students' development of algebraic thinking in the early grades, on obstacles students must overcome to reason algebraically, and on knowledge required of teachers to teach algebra well. That is, the vast majority of research on students' algebra has focused on learning and teaching algebra. Very little research has focused on the various consequences for students' later mathematical learning of the algebra that they actually learn. This book picks up that focus. It addresses the consequences of the algebra students learn in school for their mathematical learning in college. It also addresses a problem created for students by college instructors' attitude that, in solving mathematical problems or proving theorems, "We've done the hard, strategic part. *The rest is just algebra.*" It turns out that "the rest," for students, is not just algebra. It too often is an abyss they fall into with great apprehension and anxiety, an abyss that distracts them from strategy and meaning and obstructs their insight into larger themes.

Actually, this book does more than trace consequences of students' difficulties with school algebra. Chapters in it offer new insights into

- Sources of students' difficulties in algebra (e.g., impoverished understandings of fraction and proportion)
- The interplay between affect and mathematical understanding
- The role of reflection in students' successful algebra learning
- Alternative conceptions of algebra (e.g., emphasis on functions and modeling)
- The centrality of generalizing and particularizing in algebraic thinking
- Critical ways of thinking that curricula and instruction fail to foster (e.g., variables vary and expressions have numerical values)
- Uses of computer algebra systems to help students develop symbol sense and structure sense while, ironically, removing any need that they engage directly in algebraic manipulations

- New ways to think about linear algebra that both deepen and broaden students' school algebra.

Put another way, this book contains a collection of chapters that put past research on students' algebra in a new light. It also offers new ways to think about addressing students' difficulties in school algebra while at the same time offering ways to envision how we might support students' meaningful uses of algebra beyond school mathematics.

Tempe, AZ
June 2016

Patrick W. Thompson

Introduction

In solving mathematics problems in college level, very often one reaches a point where certain algebraic manipulation is a required and unavoidable part of the procedure. Once a mathematician reaches this point the rest is trivial, straightforward, and in some sense a pleasant completion to a long process. Namely, the real mathematics part is taken care of "and the rest is just algebra," an expression commonly used by many mathematics instructors. On the contrary, many college students undergo an entirely different experience. These students often halfheartedly apply the more advanced theories and make some progress in a problem-solving situation; however, the algebra portions which are initially hidden or mixed in with other contexts create an enormous obstacle. The unresolved high school algebra knowledge is analogous to a *tornado wrapped in rain*, appearing in so many courses where outwardly seem far removed from high school algebra. In college as the complexity of mathematical ideas increases rapidly, the unresolved high school algebra problems mount up progressively and continue to create further distress. While assessing students' work, it is often difficult to unravel their thought processes and the convoluted algebraic errors which are challenging to accurately categorize or justify.

Increasing complexity ↓

$$\frac{1}{2}+\frac{1}{16}$$

$$\frac{1}{x^2}+\frac{1}{x^4}=50$$

$$\frac{\mathrm{d}}{\mathrm{dx}}\left(\frac{x^2}{16}\right)$$

$$\lim_{x\to 0}\frac{\sqrt{1+x^2}-\sqrt{1-x^2}}{x}$$

Many college instructors are facing this dilemma every day. Students who seemingly follow more complex mathematical concepts are unable to proceed as problems, for example involving fractions, will soon let them down. In college level, students who cannot perform what is known as "basic algebraic manipulations," or do not possess adequate "assumed algebra knowledge" readily available, face significant disadvantages. This cohort of students are often not able to follow the instructors' problem-solving steps and easily get lost if any steps are skipped. They are often unable to successfully reach a desired solution on their own, causing frustrations for themselves, their instructors, and consequently raising many concerns for the institution and the country as a whole.

This book is a coordinated collection of chapters written by several experts in the field that addresses one of the most persistent mathematics pedagogy challenges of this century. The authors, who have critically examined students' difficulties from their areas of research, emphasize that these difficulties are more complex than just forgotten rules and offer strategic approaches that hold promise of greater success for more college students. Their research and discussion will raise awareness on the complexity and challenges facing the mathematics community. Mathematics instructors who are frustrated with their students' lack of skills and knowledge will find this volume helpful, as the authors confront the question of why students have difficulties with algebra and reveal how to improve their long-term understanding and success.

The first part of the book brings issues regarding the current state of students' inadequacies and fluency of algebra in the US colleges and is divided into two chapters.

In the first chapter, Stewart and Reeder set the scene by showing how the unresolved high school algebra misconceptions and shortcomings create major complications in college mathematics courses. The examples that are illustrated in this chapter are the tip of the iceberg, showing the types that most instructors will frequently come across in assessing students' work. The consequences of lacking solid algebra in college level are devastating and deserve to be addressed appropriately.

The second chapter by McGowen shows how the problematic nature of prior knowledge hinders students' success in college-level mathematics. She cites the need to improve the effectiveness of teaching and calls for teachers to be aware of students' problematic met-befores, providing guidance accordingly, in order to help students develop deeper understanding of mathematics and promote mathematical thinking.

The second part of the book, devoted to algebra in a broader context, is divided into three chapters.

In the first chapter, Tall offers a combined framework for mathematics in general and algebra in particular and contemplates why some students find algebra pleasurable whereas others find it a source of anxiety. In his view, many students are affected by problematic aspects that have accumulated over many years and become more difficult to address as the ideas become deeply ingrained. Tall suggests a significant re-think in how we view the development of mathematical thinking that promotes flexible thinking on the one hand and impedes long-term learning on the other.

The second chapter presented by Booth, McGinn, Barbieri, and Young is concerned about students' common misconceptions. The authors believe that traditional instruction will not remedy the problem. Drawing from a wealth of mathematics education and educational psychology literature, the authors offer a number of interventions to address these misconceptions. Furthermore, they discuss ways of preventing them from developing.

The third chapter by Reeder highlights the significance of mathematics teachers' content and pedagogical knowledge. In her views, to prepare successful students who are able to contribute to the global economy, we must insist that the mastry of basic skills is not suffecient. In her views, the process of preparing students well is challenging and requires a holistic approach, but can be met.

The third part of the book, devoted to positive approaches to the teaching of algebra, is divided into two chapters.

Drawing on years of experience working with many students from all ages, Mason believes that once the learner appreciates where the algebraic expressions come from, manipulating them is a straightforward task. This chapter offers well-thought out pedagogical strategies, didactic tactics, and specifically designed tasks and offers ways forward on how to succeed in learning algebra.

Fey and Smith suggest a bold curriculum change, centering on functions. They assert that the current US high school curriculum has no room for more applied mathematics. Topics such as probability, statistics, modern discrete mathematics, and mathematical modeling must be included to prepare our students to solve real life problems. In this chapter, the authors assess their proposed curriculum by carefully considering the challenges and responding to them.

Combining historical and current didactical ideas, Nataraj and Thomas reveal fresh approaches to the understanding of algebra. The authors discuss the various uses of letters and the concept of exponentiation and emphasize that in order to prepare students for long-term understanding and success, much groundwork needs to be established in the middle and lower secondary years of schooling.

The fourth part of the book proposes future developments and is divided into two chapters.

Drawing from studies in mathematics educational neuroscience, Kieran reveals new findings offering different insights into how we think about the so-called just algebra part of a mathematical problem. Kieran encourages the mathematics community to reflect and question their traditional beliefs that assume algebraic activities are straightforward algorithmic procedures.

The second chapter presented by Thomas examines how particular tasks, including some that integrate digital technology into student activity, could be used to re-think the algebra curriculum content with a view to motivating students and promoting versatile thinking. Thomas finds the underlying principles of Framework of Advanced Mathematical Thinking (FAMT) prove to be useful in school level mathematics.

The last part of the book, dedicated to teaching higher algebra, is divided into two chapters.

To prepare students to embrace abstraction, Hannah recommends allowing linear algebra to evolve naturally from students' experiences. He believes provoking a need for definitions and theorems is a far better approach than presenting a set of predetermined fully formed axioms.

Stewart believes that the sudden and to some degree unexpected entry into more formal thinking in linear algebra causes difficulties for many undergraduates who do not have a prior background. Based on the FAMT, Stewart demonstrates the type of thinking that is expected and required in college level in order to succeed. She recommends these types of thinking ought to be introduced and encouraged in high school.

The purpose of this book is not to offer a magic remedy or a set of polished guidelines to quickly resolve a deeply rooted problem. The intention of this book is to reach out to the mathematics community and encourage every person to take a bold step toward changing how we teach and learn algebra. Talking about what our college students are not capable of doing and essentially sitting on the fence and watching it all go by is no longer an option. We all have some appreciation of what our students can or cannot do. The question is what are our responsibilities in regard to college students' poor algebra knowledge?

Department of Mathematics
University of Oklahoma
Norman, OK
USA

Sepideh Stewart

Acknowledgments

I would like to express my sincere gratitude to each author for embracing the theme of this book and producing such high-quality and thought-provoking work. It has been an honor and privilege for me to work with a team of outstanding authors. My special thanks goes to David Tall and Mercedes McGowen for their invaluable advice and guidance throughout this project.

I would also like to thank Dr. Andy Miller, who was the Chair of the Department of Mathematics at the University of Oklahoma during the time this book was being prepared, for his encouragement and support.

I would like to extend my appreciation to Melissa James, Springer's Publishing Editor, for her continuous support and indispensable knowledge.

Norman, OK Sepideh Stewart

Contents

About the Authors

Christina Barbieri is an Institute of Education Sciences (IES) Postdoctoral Research Fellow at the University of Delaware, where she works with Nancy Jordan, Roberta Golinkoff, and Henry May on research that focuses on bridging the gap between cognitive and education science. She received her B.A. in Psychology from the City University of New York and then went on to receive her Ph. D. in Educational Psychology at Temple University. There, her research with Dr. Julie Booth focused on exploring algebra learning with consideration of cognitive principles. Christina's primary line of research focuses on instructional strategies that can potentially reduce students' misconceptions about math and number.

Julie L. Booth is an Associate Professor of Educational Psychology at Temple University, where she also currently serves as the Associate Dean of Undergraduate Education for the College of Education. She received her Ph.D. in Psychology in 2005 from Carnegie Mellon University under Dr. Robert Siegler, and trained as a postdoctoral fellow at the NSFfunded Pittsburgh Science of Learning Center, where she conducted research on students' learning in real-world classrooms with Dr. Ken Koedinger. Dr. Booth has received funding from both IES and NSF as PI or co-PI on 8 federal grants, including the National Center for Cognition and Mathematics Instruction led by WestEd and the AlgebraByExample and MathByExample projects managed by the Strategic Education Research Partnership. Her research interests lie in translating between cognitive science/cognitive development and education by finding ways to bring laboratory-tested cognitive principles to real-world classrooms, identifying prerequisite skills and knowledge necessary for learning, and examining individual differences in the effectiveness of instructional techniques based on learner characteristics.

James T. Fey is Professor Emeritus in the Department of Mathematics and the Department of Teaching and Learning, Policy and Leadership at the University of Maryland. For four decades, the focus of his professional scholarship has been a series of projects that have designed, developed, and tested innovative curricula and instructional materials for middle and high school mathematics, with special concentration on algebra. He was project director for the NSF-funded Computer Intensive Algebra project, and he has been a principal investigator on the NSF-Funded Connected Mathematics Project and the Core-Plus Mathematics Project since their inception. Those projects emphasized development of curriculum materials to support problem-based instruction, especially the kind of functions-oriented approach to algebra that will be the centerpiece of this new project. They resulted in publication of widely used instructional materials, empirical research on efficacy of the innovations, and theoretical essays on mathematics curriculum and instruction. In addition to his administrative and curriculum development experience from the three major projects, Dr. Fey was project director for the NSF-funded Maryland Collaborative for Teacher Preparation and the Mid-Atlantic Center for Teaching and Learning. From 2008 to 2010 he completed a 2-year rotation as a visiting scientist in the Division of Research on Learning in Formal and Informal Settings at NSF where his program officer responsibilities focused on review, award, and monitoring of grants for projects in the Discovery Research K-12 program.

John Hannah is a Senior Lecturer at the University of Canterbury in Christchurch, New Zealand. He did his Ph.D. in Ring Theory and published a series of papers (often with K.C. O'Meara) about von Neumann regular rings. More recently he has become interested in mathematics education, especially as it relates to visualization and the learning of linear algebra, and in the history of mathematics, especially the role of visual imagery in the development of early algebra.

Carolyn Kieran is Professor Emerita of Mathematics Education at the Université du Québec à Montréal where her university career began in 1983 in the Department of Mathematics. Her primary research interest is the learning and teaching of algebra, with a particular focus on the roles played by computing technology. She served as President of the International Group for the Psychology of Mathematics Education from 1992 to 1995 and as member of the Board of Directors of the National Council of Teachers of Mathematics from 2001 to 2004. She is currently Chair of the Klein and Freudenthal Awards Committee of the International Commission on Mathematical Instruction. Past contributions to the research community include the International Program Committee for the 22nd ICMI Study on Task Design in Mathematics Education and the Advisory Boards of the First and Second Handbooks of Research on Mathematics Teaching and Learning. Recent publications comprise the Algebra Teaching and Learning entry in the *Encyclopedia of Mathematics Education* and chapters in the *Third International Handbook of Mathematics Education, Early Algebraization: A Global Dialogue from Multiple Perspectives,* and *Vital Directions in Mathematics Education Research.*

John Mason has been teaching mathematics ever since he was asked to tutor a fellow student when he was fifteen. In college he was at first unofficial tutor, then later an official tutor for mathematics students in the years behind him, while tutoring school students as well. After a B.Sc. at Trinity College, Toronto, in Mathematics and an M.Sc. at Massey College, Toronto, he went to Madison Wisconsin where he encountered Polya's film "Let Us Teach Guessing," and completed a Ph.D. in Combinatorial Geometry. The film released a style of teaching he had experienced at high school from his mathematics teacher Geoff Steel, and his teaching changed overnight.

His first appointment was at the Open University, which involved among other things the design and implementation of the first mathematics summer school (5000 students over 11 weeks on three sites in parallel). He called upon his experience of being taught to institute active-problem-solving sessions, which later became investigations. He also developed project work for students in their second year of pure mathematics. In 1982 he wrote *Thinking Mathematically* with Leone Burton and Kaye Stacey, which has turned into a classic (translated into four languages; now in an extended new edition), and is still in use in many countries around the world with advanced high school students, with graduates becoming school teachers, and with undergraduates in courses in which students are invited to think about the nature of doing and learning mathematics. *Learning and Doing Mathematics* was originally written for Open University students, then modified for students entering university generally.

At the Open University he led the Centre for Mathematics Education in various capacities for 15 years, which produced the influential *Routes-to Roots-of Algebra*, and numerous collections of materials for teachers at every level. His principal focus is thinking about mathematical problems and supporting others who wish to foster and sustain their own thinking and the thinking of others. Other interests include the study of how authors have expressed to students their awareness of generality, especially in textbooks on the boundary between arithmetic and algebra, and ways of working on and with mental imagery in teaching mathematics. *Mathematics Teaching Practice: a guide for university and college lecturers* is a distillation of over one hundred tactics for informing the teaching of mathematics. *Fundamental Constructs in Mathematics Education* is a collection of extracts from research literature, intended for masters students seeking entry into the complex world of mathematics education research. *Designing and Using Mathematical Tasks* brings together the various frameworks used in mathematics education courses at the Open University over 25 years. In *Practitioner Research Using the Discipline of Noticing* he has articulated a way of working developed at the Centre which provides methods and an epistemologically well-founded basis for practitioners to develop their own practice, and to turn that into research. With his wife Professor Anne Watson at the University of Oxford he has tried to support the teaching of mathematics and the development of mathematical thinking through publications such as *Questions and Prompts for Mathematical Thinking*; *Thinkers*; and *Mathematics as a Constructive Activity: learners generating examples.*

Having retired in 2009, John is professor emeritus at the Open University and Senior Research Fellow in the Department of Education at the University of Oxford. He continues to work with people who want to foster and sustain mathematical thinking in themselves and in others.

Kelly M. McGinn is a postdoctoral fellow working on the IES-funded Math by Example project with Dr. Julie Booth. She holds a B.A. in Psychology from the Catholic University of America, a certification in both secondary general science and middle school mathematics from the University of Pennsylvania, and a Ph.D. in Educational Psychology from Temple University. Her research focuses on the development of mathematics conceptual understanding and is especially interested in the relation between the use of mathematical language and conceptual development.

Mercedes McGowen is Professor Emerita of Mathematics at William Rainey Harper College in Palatine, IL. Her major responsibilities included curriculum development and teaching pre-service teacher mathematics courses and developmental algebra. The teacher preparation program she developed and two mathematics textbooks she coauthored (*Applying Algebraic Thinking to Data* and *Mathematical Investigations: Concepts and Processes for the Introductory Algebra Student*) received Exemplary Program awards from the National Science Foundation, American Mathematics Association of Two-Year Colleges, and the Illinois Community College Board. Prior to teaching at William Rainey Harper College, Mercedes taught junior high and high school mathematics.

Her primary research interest is the learning of algebra by undergraduates enrolled in developmental courses and the teaching of these courses, with a particular focus on growth of flexible thinking. Currently, she is working with the University of the Nouvelle Grande Anse, Jeremie, Haiti (UNOGA), coauthoring materials for a Divergent Thinking Seminar (2014) that the university intends to incorporate as a required course for all UNOGA students. She also coauthored the materials for a mathematics course at the university (2012). Recent publications include *Flexible thinking and met-befores: Impact on learning mathematics*; *Metaphor or Met-before? The effects of previous experience on the practice and theory of learning mathematics*; and *Changing Pre-Service Elementary Teachers' Attitudes to Algebra*. Past contributions to the science of teaching and learning include the College Board of Mathematical Sciences National Summit on Teacher Preparation, Washington, D.C. Steering Committee, and the National Research Council Mathematics Learning Study II Planning Committee.

Mala S. Nataraj is a high school mathematics teacher and teaches at Selwyn College, Auckland, New Zealand. She has completed her Ph.D. in mathematics education and continues to research alongside teaching. Mala is passionate about mathematics and teaching and is committed to pedagogical practices that enhance mathematical understanding.

Stacy Reeder is Associate Professor of Mathematics Education and Chair of the Department of Instructional Leadership and Academic Curriculum. Her research interests include rational number concepts, mathematics teacher education, and problem-centered learning. Stacy is a former high school mathematics teacher and current Chair of the Mathematics Education Graduate and Undergraduate programs as well as the Elementary Mathematics Specialist Certification program at OU. In 2014 she was awarded the JRCoEd Teaching & Advising award and received the Jon E. Pedersen Award for Excellence in Mentoring Graduate Students.

David A. Smith is Associate Professor Emeritus in the Department of Mathematics at Duke University and a Fellow of both the American Mathematical Society and the American Association for the Advancement of Science. In addition to more than 40 years service at Duke, he taught during sabbatical years at Benedict College, Case Western Reserve University, and Pennsylvania State University. He is the author or coauthor of two college-level textbooks and many other instructional materials, as well as about a hundred published papers in abstract algebra, combinatorics, mathematical psychology, numerical analysis, and mathematics education. He was the founding editor of the *Journal of Online Mathematics and its Applications* and has served on a number of other editorial boards and advisory committees, mostly for the Mathematical Association of America. He was Co-Director (with Lawrence Moore) of *Project CALC: Calculus As a Laboratory Course*, an award-winning NSF-funded curriculum development project. From the inception of the Core-Plus Mathematics Project, he has served on its National Advisory Panel.

Sepideh Stewart is an Assistant Professor of Mathematics Education in the Department of Mathematics at the University of Oklahoma. Her research interests are on the embodied symbolic and formal worlds of mathematical thinking with a particular interest on the pedagogy of linear algebra. Over the past 4 years she has also been working on several projects investigating the mind of working mathematicians. Her latest research project on algebra is examining undergraduate students' difficulties with school algebra, which are causing major tensions in the calculus courses.

David Tall is Emeritus Professor in Mathematical Thinking at the University of Warwick. He studied for his doctorate in mathematics with Fields Medalist Sir Michael Atiyah and his doctorate in the psychology of mathematical thinking with Richard Skemp. He has published books on *Foundations of Mathematics*, *Complex Analysis*, and *Algebraic Number Theory* with Ian Stewart, designed and programmed *a Graphic Approach to the Calculus* blending dynamic visualization and operational symbolism, and was seminal in developing research on mathematics learning at university level including the book *Advanced Mathematical Thinking*. His latest book on *How Humans Learn to Think Mathematically* offers an overall framework for the cognitive and affective development of mathematical thinking from child to adult that also relates directly to the historical evolution of mathematics.

Mike Thomas is Emeritus Professor of Mathematics Education in the Mathematics Department at The University of Auckland, New Zealand. He is interested in semiotic analyses of gestures and representations in learning mathematics, including links to the nature of advanced mathematical thinking, the learning and teaching of algebra and calculus, and the transition from school to university and undergraduate mathematics, such as linear algebra. He also investigates the role of teacher affective factors, such as orientations and confidence, in school and university teaching and professional development, often with an emphasis on the use of digital technology. He has given invited research seminars in a number of countries and is on the editorial boards of the Springer international journals, *Mathematics Education Research Journal* and *Digital Experiences in Mathematics Education*. He is Editor-in-Chief of the *International Journal of Research in Undergraduate Mathematics Education*, published by Springer. He led a survey team for the 2012 International Congress on Mathematical Education (ICME), on the mathematical difficulties inherent in the transition from school to university.

Laura K. Young is a graduate research assistant and student in Temple University's Educational Psychology Ph.D. program. Her undergraduate degree is in Human Development and Family Studies from Penn State University, Brandywine. Her research explores the development of mathematical skills in both formal and informal contexts, how these skills change developmentally, and whether they can predict later mathematics achievement. Additionally, her research examines the use of interventions, such as worked examples, all in an attempt to understand and improve mathematical abilities.

Part I
Current State of Students' Inadequacies, Conceptions and Fluency of Algebra in the US Colleges

Chapter 1
Algebra Underperformances at College Level: What Are the Consequences?

Sepideh Stewart and Stacy Reeder

> *...Although I couldn't really prove a lot on the exams, I did learn how to solve calculus problems, unfortunately what held me back was the algebra.*
>
> —Calculus I Student

Abstract Many college instructors consider the final problem-solving steps in their respective disciplines as "just algebra"; however, for many college students, a weak foundation in algebra seems to be a source of significant struggle with solving a variety of mathematics problems. The purpose of this chapter is to reveal some typical algebra errors that subsequently plague students' abilities to succeed in higher-level mathematics courses. The early detection and mindfulness of these errors will aid in the creation of a model for intervention that is specifically designed for students' needs in each course.

Keywords Algebra • Common errors • Undergraduate mathematics • Student difficulties, calculus, college mathematics

1.1 Introduction

As mathematics instructors we deal with students' algebra shortcomings on a day-to-day basis. In reality, in many problem-solving scenarios, after applying a new theory, algebra takes over the rest of the process. While many students seemingly follow the theories that are introduced in mathematics lectures in college-level courses, in many cases not having a rich algebra machinery, including some level of the mastery of the algebraic symbols and expressions, prevents them from

---

S. Stewart (✉)
Department of Mathematics, University of Oklahoma, 601 Elm Ave, Norman, OK 73019, USA
e-mail: sstewart@math.ou.edu

S. Reeder
University of Oklahoma, 820 Van Vleet Oval, ECH 114, Norman, OK 73019, USA
e-mail: reeder@ou.edu

S. Stewart (ed.), *And the Rest is Just Algebra*, DOI 10.1007/978-3-319-45053-7_1

completing basic tasks. The hierarchical nature of mathematics becomes particularly apparent when the algebra component gets in the way and does not play the role in the manner it should. Many students become particularly frustrated as they soon discover that the fast pace of college mathematics lectures and new material is not going to wait for them to catch up. On the other hand, the instructors become disappointed with students' performances as they see instances of algebra errors that should not have entered with students into the university and should have been resolved years ago in high school. Astonishingly, some students proceed to their senior year, still not fully grasping the fundamental aspects of mathematics.

This chapter exposes a sample of students' algebra misconceptions and their consequences, illustrating how these seemingly trivial common errors are causing significant disruption in solving problems in a variety of mathematics courses. The examples will show how algebra errors can rapidly terminate the flow of the problem-solving sequence for students and result in incorrect solutions or no solutions at all.

1.2 Research on Students' Difficulties with Algebra

Algebra is often referred to as a gateway course because it is foundational and fundamental to success in STEM subjects (Adelman, 2006), and with increased expectations on what students complete in high school, more students are taking algebra now than ever before (Stein, Kaufman, Sherman, & Hillen, 2011). Traditionally, in college, algebra content is considered as assumed knowledge, and professors are not expected to, nor have time to, reteach it. Needless to say, calculus curricula are demanding and fast moving leaving no extra time to resolve basic algebra issues.

Regrettably, research and the scores on international tests reveal that the USA has been steadily falling behind many of the other industrialized nations in terms of mathematics and science education and the production of STEM graduates. Students who lack a solid understanding of high school algebra tend to struggle in college-level calculus courses and may subsequently be deterred from pursuing STEM field degrees. A survey published by the National Center for Education Statistics reported that nationwide, in 2000, 28 % of incoming freshmen took a remedial mathematics class (US Department of Education, National Center for Education Statistics (NCES), 2004). Beyond those who find themselves underprepared for college-level mathematics coursework, the majority of students struggle in many of their college-level mathematics coursework due to incomplete or insecure understandings of many important algebraic topics. The impact of weak or incomplete mathematical understanding and algebra in particular, at the middle school and high school level, has a profound impact on the future mathematical success of students and their educational possibilities (Booth & Newton, 2012; Brown & Quinn, 2007; Wu, 2001).

More than a century ago, De Morgan (1910) wrote about the difficulties students face in learning mathematics noting common errors related to arithmetic and rational number computation. Since that time, other researchers have catalogued common errors in computation and algebra (Ashlock, 2010; Benander & Clement, 1985; Booth, Barbieri, Eyer, & Paré-Blagoev, 2014). Benander and Clement (1985) catalogued errors students made in basic arithmetic and algebra courses. Their work involved classroom observations and resulted in 11 categories of common errors including basic problem-solving skills, averages, whole numbers, fractions, decimals, percents, integers, exponents, simple equations, ratios and proportions, geometry and graphing. Ashlock's (2010) work published as a book focused on the mathematics work of school-aged children and on helping instructors thoughtfully analyse their students' work in order to discover patterns in their errors for the purpose of improving instruction. Ashlock suggests that as students learn about mathematical operations and methods of computation, they often develop and adopt misconceptions and procedural errors. Teachers who understand that this occurs and are able to identify these problems in their students' work can develop strategies to help students. In a more recent study, Booth et al. (2014) focused on the errors in algebra with school-aged students and identified errors that were "persistent and pernicious" given their predictive ability for student difficulty on standardized test items. Their study involved an in-depth analysis of students' errors during problem-solving at different points during the year and resulted in the classification of these errors which include *variable* errors, *negative sign* errors, *equality/inequality* errors, *operation* errors, *mathematical properties* errors and *fraction* errors.

Rather than classifying types of errors, Drouhard and Teppo's (2004) work focuses on the idea of *denotation* and suggests that it is a developed sense about what one is writing and a lack of sense regarding denotation creates significant problems for students. They note "that students with poor capabilities to recognize this aspect of the meaning of an expression often make endless calculations because they do not know in what direction to go and when to stop" (p. 235). Harel, Fuller and Rabin (2008) further comment on the idea of meaning and denotation indicating that students often cancel within problems without attending to the quantitative meaning of their action. For example, Harel et al. (2008) state "it is not uncommon for students to manipulate symbols without a meaningful basis that is grounded in the context in which the symbols arise; for instance a student might write: $(\log a + \log b)/\log c = (a + b)/c$" (p. 116). In this case, students may be overgeneralizing their use of the distributive property and cancel "log" without considering the quantitative meaning of their action. Harel (2007) suggests that the lack of emphasis on mathematical meaning that students, and perhaps their teachers, apply to mathematical symbols creates what is referred to as a *non-referential symbolic* way of thinking and that this way of thinking can be tied to a myriad of algebra errors. Sfard and Linchevski (1994) believe that students must be motivated "to actively struggle for meaning at every stage of learning" (p. 225). They are concerned that "if not challenged, the pupil may soon reach the point of no return, beyond which what is acceptable only as a temporary way of looking at things will freeze into a

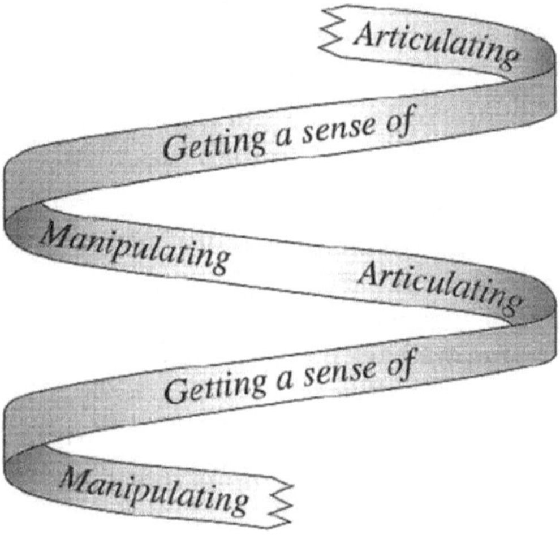

Fig. 1.1 *Manipulating-Getting-a-sense-of-Articulating* framework (Mason, 2002, p. 187)

permanent perspective" (p. 225). Mason (2002) in his framework, *Manipulating-Getting-a-sense-of-Articulating (Fig. 1.1)*, emphasizes that students must be given opportunities to make sense of situations. He believes that "students want, indeed need, confidence-inspiring familiar objects to manipulate and on which to try out new ideas so that they can literally 'make sense' of them" (p. 187). In Harel and Sowder's (2005) opinion, "instruction (or curriculum) that ignores sense-making, for example, can scarcely be expected to produce sense-making students" (p. 46).

Research in the area of student errors expresses the common belief that these errors exist in students' mathematical work and that understanding these errors and developing the ability to identify them hold promise for helping students succeed in mathematics and have implications for teaching. Further, specific interventions and strategies can be developed to help students overcome and perhaps avoid the misconceptions that underlie the errors. Booth et al. (2014) suggests that "the misconceptions underlying specific persistent errors are not corrected through typical instruction and may require additional intervention in order for students to learn correct strategies" (p. 21). Unfortunately, if these errors are not addressed and persist with students into their university-level courses, they have the potential to create significant challenges and barriers for students. Although, research on students' difficulties with algebra in school has been well documented (e.g. Hoch & Dreyfus, 2004; Kieran, 1992), in our knowledge methodical studies on the presence of these difficulties and their impact at university level are scarce. In their edited book entitled *The Future of the Teaching and Learning Algebra*, Stacey, Chick and Kendal (2004) discussed the main problems of algebra in school algebra, but again, very little was mentioned in the way of consequences for college-level mathematics.

1.3 Some Common Algebra Errors in College Mathematics Topics

1.3.1 Finding the Absolute Maximum and Minimum of a Function

Finding the absolute maximum and minimum of a function is a typical question in a first year calculus course. Stewart (2014) in his textbook, *Calculus* (8th ed.), describes the procedure as follows and ensures that the following three-step procedure will always work.

The Closed Interval Method

To find the absolute maximum and minimum values of a continuous function f on a closed interval $[a,b]$:

1. Find the values of f at the critical numbers of f in (a,b).
2. Find the values of f at the endpoints of the interval.
3. The largest of the values from steps 1 and 2 is the absolute maximum value; the smallest of these values is the absolute minimum value (p. 209).

The following example (see Fig. 1.2) demonstrates the above procedure by finding the derivative of the function $f(x)$ first. Many students are capable of applying the product rule and finding the derivative correctly. However, their struggles with algebra may prevent them from proceeding beyond this first step, as they face the challenge of simplifying the derivative function which will enable them to work with an easier form of the function. Furthermore, adding two fractions involving square roots will add to the complexity of this problem. Step 3 will also have its own challenges as it requires setting the fraction equal to zero and solving for x. Step 4 will only become possible, if the x values were deduced successfully from the previous steps.

The following examples of work by students show common errors among many (see Table 1.1). Although, the work by students varies, typically they all have one thing in common, lack of fluency with algebra. Students (1) and (2) tried to avoid fractions and dealing with square roots. Although student (2) had some knowledge of the calculus and managed to write the algorithm in a box as a reminder, her lack of fluency with algebra eventually let her down. Her work revealed that she was guessing solutions for x and got wrong critical numbers. Although, both students (3) and (4) tried to work with the fractions, small errors along the way slowly crept in, and they too ended up with incorrect critical numbers.

1. Find the absolute maximum and absolute minimum values of f.

$$f(x) = x\sqrt{x - x^2}$$

Solution:
Step 1: Using the product rule we find the derivative of the function as follow

$$f'(x) = x\frac{1 - 2x}{2\sqrt{x - x^2}} + \sqrt{x - x^2}$$

Step 2: Simplify the derivative equation by adding the two algebraic expressions we find:

$$= \frac{(x - 2x^2) + (2x - 2x^2)}{2\sqrt{x - x^2}} = \frac{3x - 4x^2}{2\sqrt{x - x^2}}$$

Step 3: To find the critical numbers, set $f'(x) = 0$.

$$f'(x) = \frac{3x - 4x^2}{2\sqrt{x - x^2}} = 0$$

$\Rightarrow 3x - 4x^2 = 0 \Rightarrow x(3 - 4x) = 0 \Rightarrow x = 0$ or $x = 3/4$.

Step 4: Test the boundaries to find the absolute max and min
$f(0) = f(1) = 0$ absolute minimum
$f(\frac{3}{4}) = \frac{3}{4}\sqrt{\frac{3}{4} - (\frac{3}{4})^2} = \frac{3}{4}\sqrt{\frac{3}{16}} = \frac{3\sqrt{3}}{16}$ absolute maximum.

Fig. 1.2 Finding the absolute maximum and minimum of a function

1.3.2 Dealing with Algebra While Solving Limits

Research in mathematics education reveals that students frequently struggle dealing with limits (Cornu, 1991; Oehrtman, 2002; Williams, 1991). The concept is often taught graphically as well as numerically using tables. In the meantime, the discussions of limit laws and definitions related to the left and right hand limit are also introduced. Moreover, some instructors teach the precise definition of the limit using the epsilon and delta notions. However, in many problem-solving situations where limits are present, algebraic manipulations are often requested and preferred by the instructor. A sample of common errors from first year calculus students is shown in Table 1.2. The examples reveal numerous errors especially with fractions (students (5) and (7)). Student (6) avoided dealing with the square root by unsuccessfully using the numerical method.

Table 1.1 Calculus students' work on finding the absolute maximum and minimum of the above function

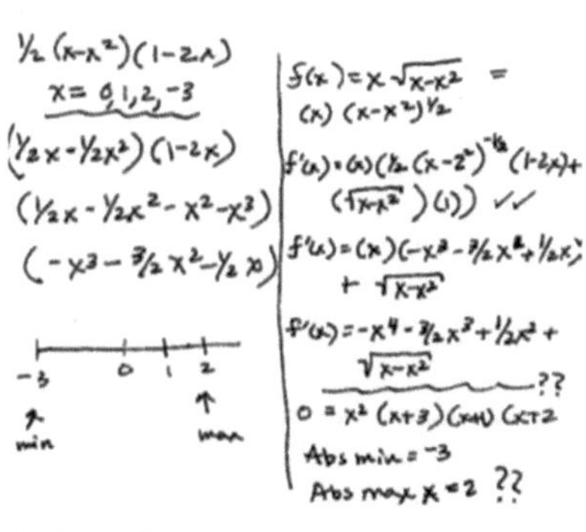 Student (1)	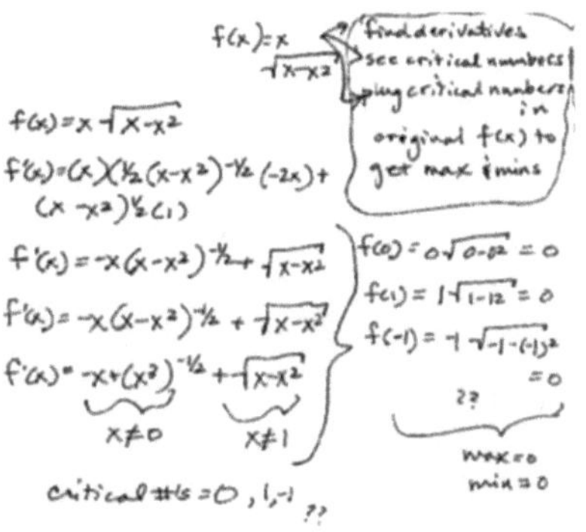 Student (2)
 Student (3)	 Student (4)

1.3.3 *Cancellation of Symbols*

Cancellation creates numerous problems in many different mathematics courses. The following table illustrates a range of student works from carelessly cancelling coefficients and symbols to cancelling almost everything (student (12)). The examples reveal interruptions in many situations involving implicit differentiations, finding the domain or integrating the function (Table 1.3).

Table 1.2 Students' difficulties with algebra while solving limit problems

 Student (5)	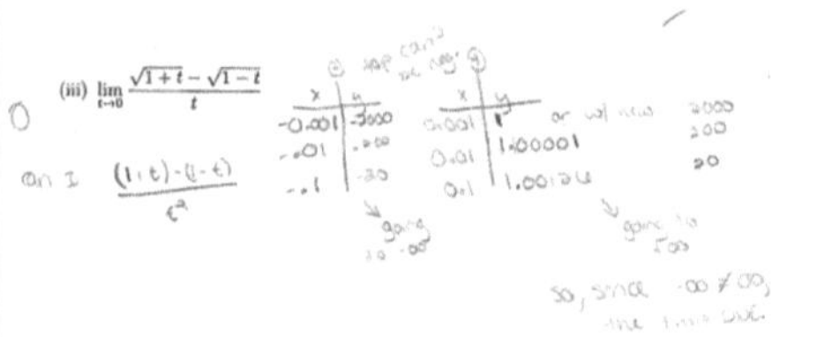 Student (6)
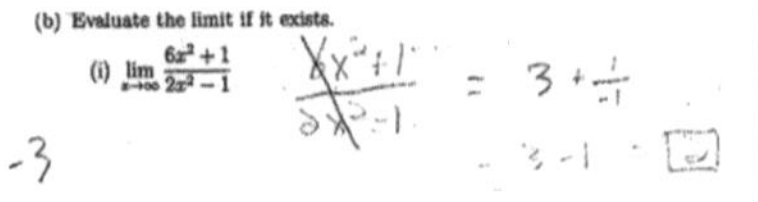 Student (7)	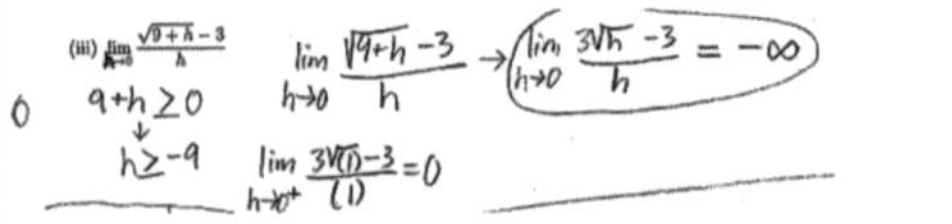 Student (8)
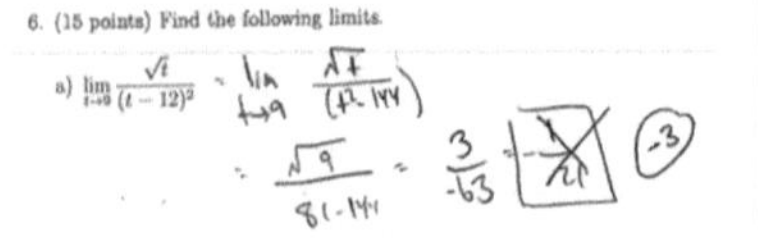 Student (9)	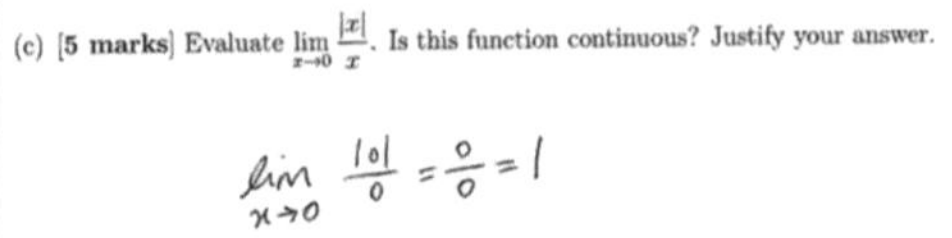 Student (10)

1.3.4 Other Algebra Errors

While looking over students' work from a first year calculus course, we noticed that one of the students (see Fig. 1.3) by default was including $x = 0$ in every case as she was going through the problems.

1.3.5 Algebra Errors in Linear Algebra

Based on the definition, a subspace is a subset of a larger space that has the following properties: (a) includes the zero vector, (b) is closed under addition and (c) is closed under scalar multiplication. For example, the vector $(a,b,1)$ for a, b belonging to any real numbers is not a subspace since it does not include the zero vector.

The following example (Fig. 1.4) is similar to examples of its kind but involves a small twist.

Each time the first author poses this question in a test, half the class loses points because of this small algebra mistake that enters into the calculation. Many students make the classic mistake of putting $(a+c)^2 = a^2 + c^2$. Here is a sample of correct (Fig. 1.5a) and incorrect (Fig. 1.5b) responses to this question.

Table 1.3 Sample of students' common errors with cancelling algebraic expressions in fractions causing significant disruption in solving problems in variety of calculus courses

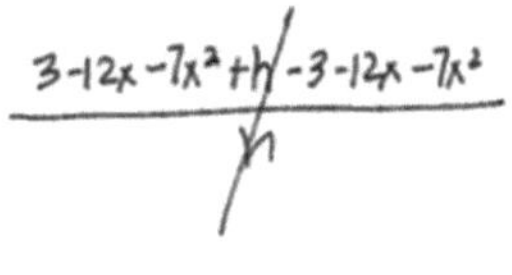 Student (11)	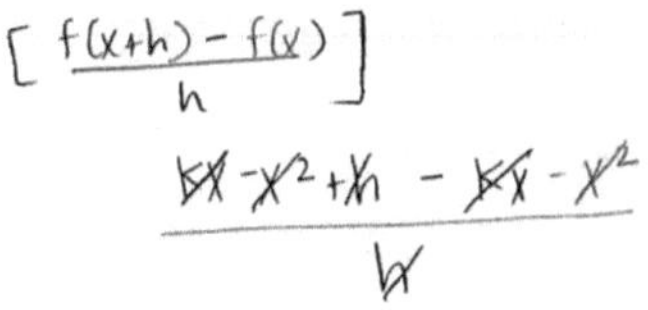 Student (12)
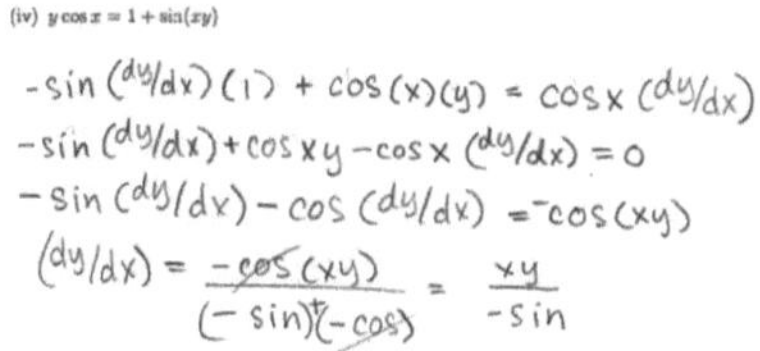 Student (13)	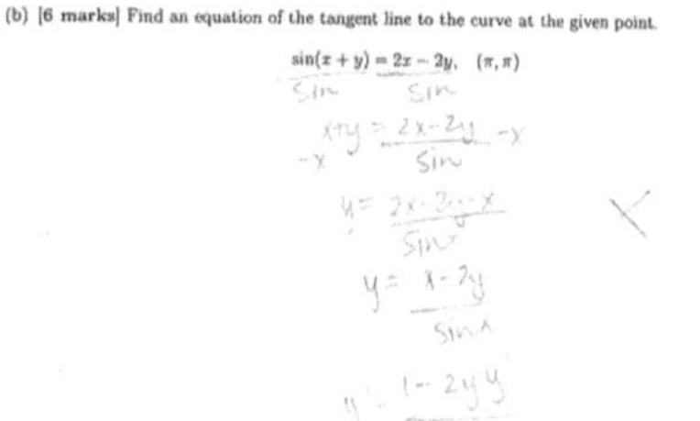 Student (14)
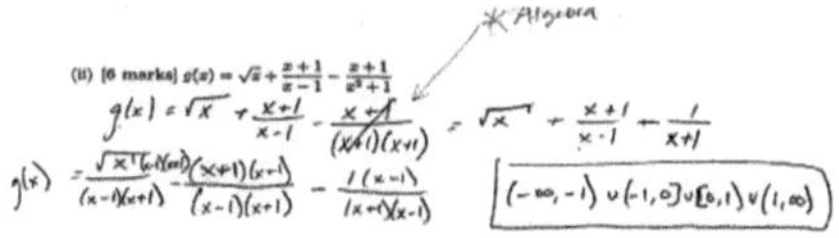 Student (15)	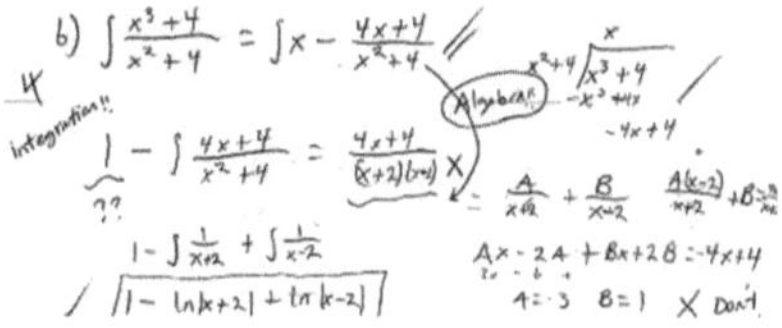 Student (16)

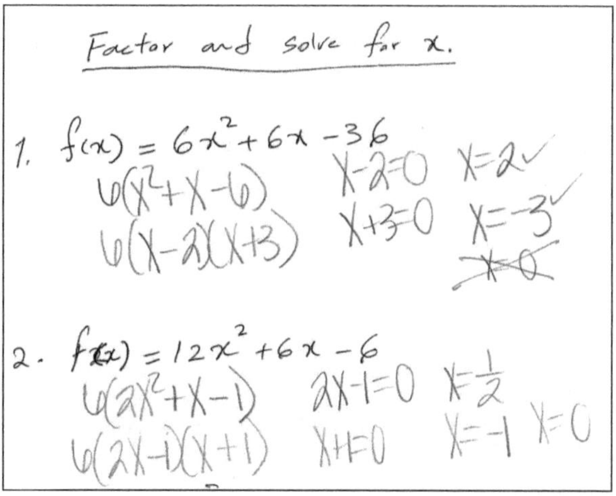

Fig. 1.3 Student's difficulties with solving for *x*

2. Let W be the set of all vectors in R^3 of the form $\begin{bmatrix} a \\ a^2 \\ b \end{bmatrix}$, where a and b are any real numbers.

Determine if W is a subspace of R^3.

Solution:
Closed under addition? No.

Adding two such vectors gives: $\begin{bmatrix} a \\ a^2 \\ b \end{bmatrix} + \begin{bmatrix} c \\ c^2 \\ d \end{bmatrix} = \begin{bmatrix} a+c \\ a^2+c^2 \\ b+d \end{bmatrix}$ Since the second component of this vector is not the square of the first $(a+c)^2 \neq (a^2+c^2)$, the vector is not in W. Therefore, W is not closed under addition.

Although this is sufficient to show that W is not a subspace, we can also show that W is not closed under scalar multiplication.

$k(a, a^2, b) = (ka, ka^2, kb)$.

Again, the first component is not the square of the second. Thus, it is not in W.

Fig. 1.4 Determining if W is a subspace of R^3

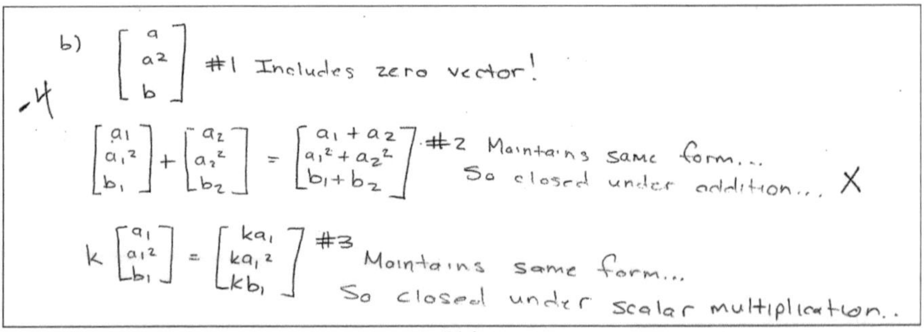

5(a)

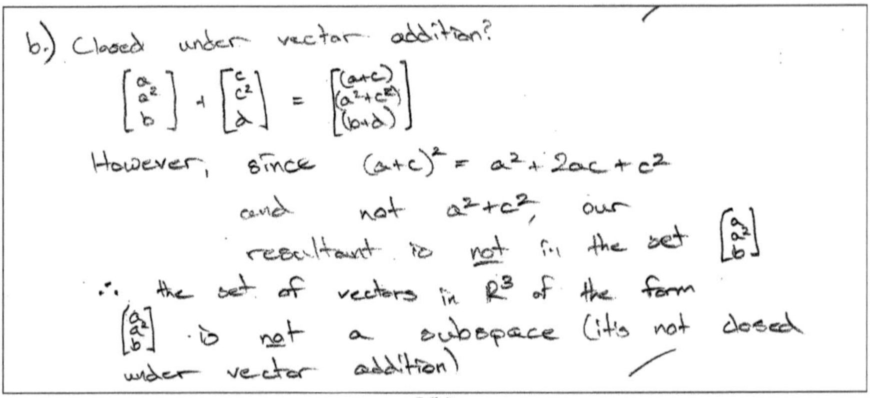

5(b)

Fig. 1.5 Linear algebra students' incorrect (**a**) and correct (**b**) responses to the question related to subspaces

1.3.6 Algebra Errors in Statistics

The algebra errors not only can occur in the lower mathematics courses, they can follow students into their later years of college. The following (see Fig. 1.6) is an example from a final exam in a senior-level statistics course.

To find the mean *E*(*Y*), the student (see Fig. 1.7) used the correct formula and successfully substituted the limits for the integration and progressed well up to the second line. At the third line, a minus sign was incorrectly introduced (first error), either as he was working or was introduced later in the process. At the fourth line, a number of algebraic errors were made as the student tried to force the situation to arrive at the given result. He seemed to have attempted factoring -1 out of $(b^2 - a^2)$ to arrive at $(b^2 + a^2)$. While not factoring the difference in squares, he made the

8. The uniform distribution has density function $f(y) = \frac{1}{b-a}$ when $a \le y \le b$ and 0 everywhere else. Prove that the mean of the distribution is:

$$E(Y) = \frac{a+b}{2}$$

then prove the variance is:

$$Var(Y) = \frac{(b-a)^2}{12}$$

Showing the derivation line by line is essential to obtain full marks. You may use the following definitions

$$Var(Y) = E[(Y - E(Y))^2]$$

$$E(Y) = \int_{-\infty}^{\infty} y f(y) dy$$

$$\begin{aligned}
E(Y) &= \int_{-\infty}^{+\infty} y f(y) dy \\
&= \int_a^b y \frac{1}{b-a} dy \\
&= \frac{1}{b-a} \left[\frac{1}{2} y^2\right]_a^b \\
&= \frac{1}{b-a} \frac{1}{2} \left[y^2\right]_a^b \\
&= \frac{1}{b-a} \frac{1}{2} \left[b^2 - a^2\right] \\
&= \frac{1}{b-a} \frac{1}{2} (b+a)(b-a) \\
&= \frac{b+a}{2}
\end{aligned}$$

$$\begin{aligned}
E(Y^2) &= \int_{-\infty}^{+\infty} y^2 f(y) dy \\
&= \int_a^b y^2 \frac{1}{b-a} dy \\
&= \frac{1}{b-a} \left[\frac{1}{3} y^3\right]_a^b \\
&= \frac{1}{b-a} \frac{1}{3} (b^3 - a^3) \\
&= \frac{1}{b-a} \frac{1}{3} (b-a)(b^2 + ab + a^2) \\
&= \frac{1}{3} (b^2 + ab + a^2) \\
V(Y) &= E(Y^2) - E(Y)^2 \\
&= \frac{(b^2 + ab + a^2)}{3} - \frac{(a+b)^2}{4} \\
&= \frac{4(b^2 + ab + a^2) - 3(a+b)^2}{12} \\
&= \frac{4(b^2 + ab + a^2) - 3(a^2 + 2ab + b^2)}{12} \\
&= \frac{b^2 - 2ab + a^2}{12} \\
V(Y) &= \frac{(b-a)^2}{12}
\end{aligned}$$

Fig. 1.6 A typical question in finding the mean and the variance in statistics

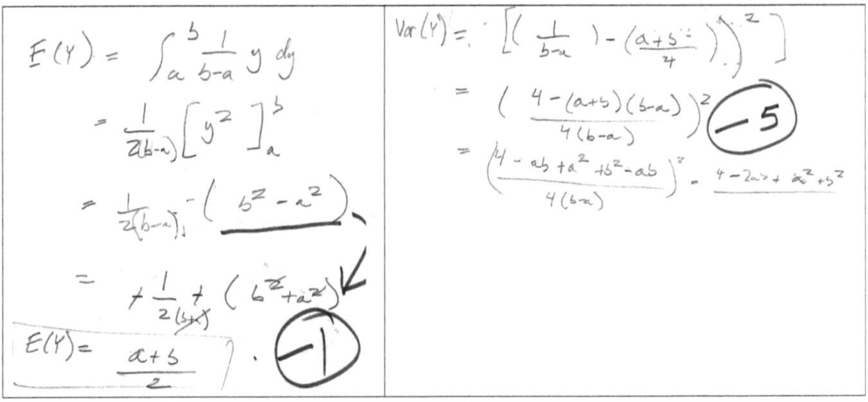

Fig. 1.7 Student's algebra errors and its consequences in the statistics final exam

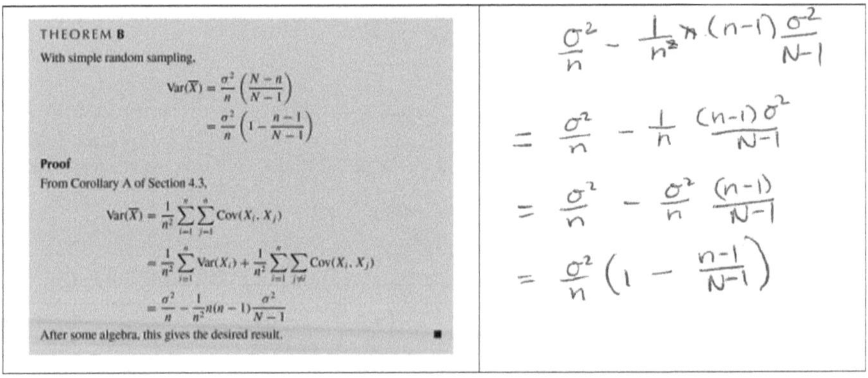

Fig. 1.8 The algebra was *left* to the reader to complete (Rice, 2007, p. 208)

classic error of cancelling $(b + a)$ with $(b^2 + a^2)$, assuming $(b^2 + a^2) = (b + a)(b + a)$, and hence forced his way to the correct result.

In finding the variance, he did not start the calculations with the correct formula $\mathrm{Var}(Y) = E\left[(Y - E(Y))^2\right]$ and carried out on the wrong path for a while and made sign errors while multiplying $-(a + b)(b - a)$ before giving up simplifying the fraction (Fig. 1.7).

1.4 Textbooks and Algebra Fluency Assumptions

It seems that the same mindset that algebra is a trivial part of a computation is also apparent in some college-level textbooks. For example, Fig. 1.8 illustrates a symbolic proof of a statistical theorem where the author has left the algebra (shown in the right) part of the proof to the reader to perform. Although, in this case it is assumed that the reader is capable of doing the algebra, many students may not make an attempt in doing the extra work and show the desired result.

1.5 Discussion

As college instructors, we encounter algebra shortcomings in our everyday interactions with students. We may express some frustration to our colleagues, at times we may feel amused as we assess students' work, and we may even blame the high school teachers or the system or the testing instrument and think to ourselves that "these students shouldn't be in my class" or, even worse, "how did they get here?" Regrettably, after a while we get used to the frequent algebra errors that our students make, and our students' challenges with algebra simply become a natural part of the landscape of our teaching. We seem to be aware of the algebra errors, even at times anticipate them, without knowing how to deal with them. It appears that saying "the rest is just algebra", habitually and unconsciously, gives us a way out in facing what is likely a complex and layered problem. Common belief among many instructors is that it is not our job to teach or reteach school algebra in our college-level courses. Realistically, going over algebra misconceptions is not a possibility, and we have no time to repair students' algebra misconceptions. Unfortunately, many of the errors presented in this chapter reflect Drouhard and Teppo's (2004) finding that students make endless calculations when they do not know what direction to go, Harel's (2007) problem with a non-referential symbolic way of thinking and the identified errors presented in Booth et al.'s (2014) study. To borrow from Booth et al. (2014), these errors are persistent and pernicious and certainly continue with students into their college-level courses. As Harel and Sowder (2005) declare, "computational shortcuts like 'move the decimal point' or 'cross multiply' or 'invert and multiply' given as rules without any attention as to why these work turns elementary school mathematics into what is deservedly called a bag of tricks" (p. 46).

Teaching new material may be complex, but restoring years of algebra misconceptions is multifaceted. Resolving the algebra deficiencies that students bring with them to college-level mathematics courses is not a trivial issue. The many connections that the students have made over the years, some based on incorrect assumptions, combined with many holes and incomplete understandings, must be carefully examined and then addressed. While this is a significant challenge and not an easy issue to address, it cannot be simply ignored and remain as an everyday accepted or out of our hands part of teaching university-level mathematics courses.

This chapter does not deal with the question of why students are making such errors and continue to make them, but rather reveals the type of errors and their consequences as well as the disruption in the flow of problem-solving actions that they cause for students. Most instructors have seen these before and in some respects nothing is new to them; however, our intention was to reveal the alarming impacts of these errors on many mathematics courses in college. We also believe that some of the problems that we observed in this chapter are more than just some forgotten rules. It is important that we as the mathematics community acknowledge the fact that these problems exist and face the challenge to seek out remedies to help the situation. The problem is complex and layered and finding a remedy cannot be achieved by looking for a temporary and quick solution. Understanding the problem and working for long-term change in the way we think, learn and teach mathematics hold promise for successfully addressing our students' challenges with algebra that plague their work, and ours, in college-level mathematics courses.

In light of the common algebra errors presented in this chapter and those documented in the literature (e.g. Booth et al., 2014) and based on our experience as mathematics instructors, we simply cannot assume our students are entering college with sufficient algebra skills and abilities. Given that we understand the devastating effect of students' weaknesses in algebra on the successful learning of college-level mathematics, it behoves us to have strategic plans to provide support and triage for them. A short-term goal would be, instead of testing students upon arrival, to provide them with a no-stake, or low-stake, formative assessment so they can assess for themselves where their weaknesses are with algebra. Tied to this assessment could be practice sets or websites with tutorials for students to review or to refresh particular concepts that will support their efforts tailored specifically for each course. Additionally, this same formative assessment could be followed up by tutoring sessions offered by the mathematics department ideally at the beginning of each semester. For example, in a college algebra or calculus course, there are always students who do not remember how to accurately factor trinomials. Students who recognize this problem via the formative assessment could attend a session to be refreshed on this topic or could work with a tutor to strengthen their problem areas early in the semester. Regardless of what intervention is developed, whether extensive or small in scope, creating an opportunity for students and their instructors to better understand the challenges students bring with them to college-level mathematics classes will help inform instruction and has the potential to help students be successful. We are aware that these short-term remedies may not be helpful to all students; however, we anticipate that they could benefit many. In Mason's (2002) views, "when difficulty arises, it is possible to retreat back down the helix (see Fig. 1.1), or even to leap down, and then to rebuild confidence and understanding while working your way back up again" (p. 188).

Beyond short-term remedies, a long-term goal will be to work closely with high school mathematics teachers as well as pre-service teacher education colleges to more effectively address the areas of concern. Creating opportunities for these groups to work together and communicate about the common errors that persist with students through their high school years and into college-level mathematics coursework will help the mathematics education community address these issues earlier. The goal of these efforts would be to help students learn important algebra concepts with depth and flexibility in their high school coursework so they do not carry with them from high school problems that will plague them for years to come and, in some cases, create barriers for pursuing careers in many STEM fields.

The authors are in the process of designing an extensive research project to address algebra errors in a variety of college mathematics courses and ultimately suggest interventions that identify, address and eliminate algebra errors. Together with Julie Booth (see in this volume), we anticipate that creating a well-thought-out model of intervention designed specifically for each course will make a positive impact in solving the algebra difficulties in university mathematics courses and will be a bold step in the right direction.

References

Adelman, C. (2006). *The toolbox revisited: Paths to degree completion from high school through college*. Washington, DC: U.S. Department of Education.

Ashlock, R. B. (2010). *Error patterns in computation: Using error patterns to improve instruction*. Boston: Allyn & Bacon.

Benander, L., & Clement, J. (1985). *Catalogue of error patterns observed in courses on basic mathematics*. Working Draft. Massachusetts: (ERIC Document Reproduction Service No. ED 287 672).

Booth, J. L., Barbieri, C., Eyer, F., & Paré-Blagoev, E. J. (2014). Persistent and pernicious errors in algebraic problem solving. *The Journal of Problem Solving, 7*(1), 3.

Booth, J. L., & Newton, K. J. (2012). Fractions: Could they really be the gatekeeper's doorman? *Contemporary Educational Psychology, 37*(4), 247–253.

Brown, G., & Quinn, R. J. (2007). Fraction proficiency and success in algebra: What does research say? *Australian Mathematics Teacher, 63*(3), 23–30.

Cornu, B. (1991). Limits. In D. Tall (Ed.), *Advanced mathematical thinking* (pp. 153–166). Boston: Kluwer.

De Morgan, A. (1910). *On the study and difficulties of mathematics*. Chicago: Open Court.

Drouhard, J. P., & Teppo, A. R. (2004). Symbols and language. In *The future of the teaching and learning of algebra the 12th ICMI study* (pp. 225–264). Boston: Kluwer.

Harel, G. (2007). The DNR system as a conceptual framework for curriculum development in instruction. In R. Lesh, J. Kaput, & E. Hamilton (Eds.), *Foundations for the future in mathematics education*. New Jersey: Erlbaum.

Harel, G., Fuller, E., & Rabin, J. (2008). Attention to meaning by algebra teachers. *The Journal of Mathematical Behavior, 27*, 116–127.

Harel, G., & Sowder, L. (2005). Advanced mathematical-thinking at any age: Its nature and its development. *Mathematical Thinking and Learning, 7*, 27–50.

Hoch, M., & Dreyfus, T. (2004). Structure sense in high school algebra: The effects of brackets. In M. J. Hoines & A. B. Fuglestad (Eds.), *Proceedings of the 28th conference of the international group for the psychology of mathematics education* (Vol. 3, pp. 49–56). Bergen, Norway: PME.

Kieran, C. (1992). The learning and teaching of school algebra. In D. A. Grouws (Ed.), *Handbook of research on mathematics teaching and learning* (pp. 390–419). New York: Macmillan.

Mason, J. (2002). *Mathematics teaching practice: A guidebook for university and college lecturers*. Chichester: Horwood.

Oehrtman, M. (2002). *Collapsing dimensions, physical limitation, and other student metaphors for limit concepts: An instrumentalist investigation into calculus students' spontaneous reasoning*, Ph.D. dissertation.

Rice, J. A. (2007). *Mathematical statistics and data analysis* (3rd ed.). California: Thomson, Brooks/Cole.

Sfard, A., & Linchevski, L. (1994). The gains and the pitfalls of reification—The case of algebra. *Educational Studies in Mathematics, 26*(2), 191–228.
Stacey, K., Chick, H., & Kendal, M. (Eds.). (2004). *The future of the teaching and learning of algebra: The 12th ICMI study* (Vol. 8). New York: Springer.
Stein, M. K., Kaufman, J. H., Sherman, M., & Hillen, A. F. (2011). Algebra a challenge at the crossroads of policy and practice. *Review of Educational Research, 81*(4), 453–492.
Stewart, J. (2014). *Calculus* (8th ed.). Boston: Cengage Learning.
U.S. Department of Education, National Center for Education Statistics. (2004). *The condition of education 2004 (NCES 2004–077).* Washington, DC: U.S. Government Printing Office.
Williams, S. (1991). Models of limit held by college calculus students. *Journal for Research in Mathematics Education, 22*, 219–236.
Wu, H. (2001). How to prepare students for algebra. *American Educator, 25*(2), 10–17.

Chapter 2
Examining the Role of Prior Experience in the Learning of Algebra

Mercedes McGowen

Abstract Widespread emphasis on developing students' algorithmic competency and symbol manipulation has resulted in students failing to think analytically and critically. If students are not encouraged to think flexibly about arithmetic and algebra in school, then this needs to be addressed by developmental courses and tasks designed to change the procedural orientation and superficial, fragmented knowledge of too many of our undergraduate students. Those who teach mathematics at the postsecondary level often dismiss the increasing number of students enrolled in precollege mathematics courses as "not my problem," not realizing that "just algebra" is the downfall for many college students. Learning "just algebra" is a much more complex task than it appears. In this chapter, prior knowledge will be shown to have become problematic for many students, and we provide evidence of the need to improve the effectiveness of our own teaching and that of our future teachers in ways that help students develop deeper understanding of mathematics and promote mathematical thinking.

Keywords Flexible thinking • Prior knowledge • Problematic met-befores • Remedial mathematics • Developmental algebra • Function machine • The minus sign

2.1 Impact of Current Instructional Practices on Student Learning

"The power of mathematical thinking—pattern recognition, generalization, problem solving, careful analysis, rigorous argument—is important for every citizen" (Barker, Bressoud, Epp, Gantert, Haver, & Pollatsek, 2004, p. 4). Mathematicians and mathematics educators claim that they want their students to think critically, make connections, and see new relationships between mathematical ideas.

M. McGowen (✉)
Department of Mathematical Sciences, William Rainey Harper College, Palatine, IL 60067, USA

601 Pleasant PL, Streamwood, IL 60107, USA
e-mail: mercmcgowen@sbcglobal.net

S. Stewart (ed.), *And the Rest is Just Algebra*, DOI 10.1007/978-3-319-45053-7_2

However, in too many classrooms, the ongoing instructional emphasis is predominantly to show students how to use a rule to get the "right" answer. The focus on mastering skills, coupled with the assumption that students understand the related mathematical concepts and terms, has failed many students, leaving them as post-graduates ill prepared for their future careers (Carlson, 1998; DeMarois, 1998; McGowen, 1998; McGowen & Tall, 2010, 2013; Oehrtman, Carlson, & Thompson, 2008; Stigler, Givvin, & Thompson, 2010; Stump, 1999).

There are those who teach mathematics at colleges and universities who dismiss the increasing numbers of students enrolled in precollege mathematics courses as "not my problem." Instructors may rightfully view what follows the initial step(s) in their respective mathematics courses as "just algebra," but in reality, "just algebra" is the downfall for many students in secondary schools and in developmental and college-level mathematics courses.

The increasing growth of undergraduate remedial mathematics courses reveals critical concerns, not only for the mathematics community but for our nation at large. Major growth in 2-year college mathematics enrollments since 1990 has been in the precollege courses (e.g., arithmetic, pre-algebra, elementary algebra, intermediate algebra, and geometry)—courses students have taken previously in elementary and high school and sometimes more than once as undergraduates. The Conference Board of the Mathematical Sciences (CBMS) (2001) survey (Blair, Kirkman, Maxwell, & American Mathematical Society, 2013, p. 136) reports that:

- In 2010, for the first time, enrollment in precollege courses at 2-year colleges totaled more than one million students (1,149,740)—a 19 % increase from 2005 to 2010.
- Arithmetic/basic mathematics, pre-algebra, and geometry course enrollments have grown from 216,000 in 2000 to 378,000 in 2010 at 2-year colleges, an increase of 75 %.
- Beginning and Intermediate Algebra enrollments increased 41 %, from 547,000 in 2000 to 772,000 in 2010.
- During the 5 years from 2005 to 2010, 4-year colleges and universities saw precollege course enrollments increase—though not as dramatically as at 2-year colleges. At 4-year colleges and universities, between 1990 and 2005, precollege course enrollment declined by 30 %.

Developmental education costs are estimated at $1 billion every year (Brothen & Wambach, 2004). Breneman and Haarlow's earlier study (1998) gave a conservative estimate of one to two billion dollars per year spent on remedial education programs at public colleges and universities. Neither of these estimates takes into account the costs incurred in time and money by students enrolled in undergraduate remedial mathematics courses. Based on analysis of data from a nationwide study of community colleges participating in the Achieving the Dream project, Bailey (2009) found that student completion rates in college English and math drop with each additional level of remedial coursework required. Only 10 % of students who placed three levels down from a college-level mathematics course pass a college-level course. Attewell, Lavin, Domina, and Levey (2006) reported that only 28 % of students who take at least one remedial course go on to complete a college credential within 8.5 years. A National Council of State Legislators article by Brenda Bautsch (2013) cites a US Department of Education study which found that only 27 % of students enrolled in remedial mathematics courses earned a bachelor's degree compared with 58 % of students who did not need remedial math. Simple "remediation" focusing on the need for accurate procedural computation does not work: at each stage, more and more students fail.

Increasing numbers of students who intend to become teachers begin their postsecondary academic careers at 2-year colleges, taking the required first 2 years of their mathematics courses at a community college before transferring to a 4-year degree program. Many preservice elementary teachers enroll in one or more developmental mathematics courses prior to taking required math content courses for preservice teachers at these institutions. Their attitudes to mathematics are generally instrumental, focused on formulas and getting correct answers. Educators of these students face a constant challenge—their students' limited understanding of what constitutes mathematics and a mathematical approach to problems.

To help students gain a more flexible, deeper understanding of arithmetic and algebra requires much more than computational or symbolic fluency. The National Council of Teachers (1989, 1991, 2000), Cohen (1995), National Research Council (2000, 2001), Conference Board of the Mathematical Sciences (2001), American Mathematical Association of Two-Year Colleges (2004), and the Common Core State Standards Initiative (2010) have all published recommendations for teachers, whether elementary, secondary, or postsecondary, directing instructors to:

- Design and implement every instructional activity guided by informed decision-making to actively engage students in the learning of mathematics
- Integrate technology appropriately into teaching to enhance students' understanding of mathematical concepts and skills
- Use results from the ongoing assessment of student learning in mathematics to improve curriculum, materials, and teaching methods

The *Curriculum Foundations Project: Voices of the Partner Disciplines* echoes these recommendations. They advocate that all teachers at every level must be able to "represent concepts in multiple ways, explain why procedures work, or recognize how two ideas are related ... be able to solve problems and to make connections among mathematical topics..." in order to modify instructional strategies and place greater emphasis on learning with understanding and focus on a thorough development of basic mathematical ideas presented in a coherent fashion (Ganter & Barker, 2004, p. xx).

Beliefs about what constitutes mathematics, what skills should be taught, when they should be taught, and to whom vary from individual to individual and community to community. Unfortunately, many mathematics departments have yet to reach a consensus acceptable to all members of the department on these issues. In the absence of mutually agreed definitions and accepted meanings among those who favor a "return to basics" and those who attempt to implement reforms in the teaching and learning of mathematics, the debate continues, with increasingly high costs to students and to our nation. Beaton (1996) cited these conflicting beliefs and practices, describing the current US mathematics curriculum as unfocused, "a splintered vision" which is reflected in our mathematics curricular intentions, textbooks, and teacher practices.

A common theme of programs described in *Models That Work* (Tucker, 1995) is that faculty in effective programs believe in "teaching for the students one has, not the students one wished one had." Unfortunately, when students lack prerequisite skills, an all-too-common reaction from instructors is "You can't expect me to reteach the entire prior curriculum. I have to teach the content of my course." This perspective serves only to block efforts to explore alternative ways of improving mathematics teaching and student learning.

The mathematics needed by first year students enrolled in many college career programs has been described as almost exclusively middle school mathematics—arithmetic, ratio, proportion, expressions, and simple equations (Common Core State Standards Initiative, 2010). Currently, instruction and learning of these topics fall far short of the understanding and competency students will be expected to demonstrate. A National Center on Education and the Economy empirical study (2013) found that introductory courses (a) fail to test complex analytical skills, the ability to synthesize materials, and solve problems not seen before as they demand only memorization of facts and mastery of procedures and (b) are not designed to test students' ability to think mathematically but instead assess memorization of facts and mastery of procedures, not the higher-order thinking skills of Bloom's taxonomy. The study also found that many community colleges have low expectations of their students, particularly in developmental courses.

The current focus on modifying technical and career programs that do not require a 4-year or advanced degree cannot result in ignoring the needs of many students who do require a deeper understanding of basic algebra concepts and skills proficiency. These students include not only our prospective elementary, secondary, and postsecondary teachers who are not currently being well-enough prepared but other non-STEM majors who also need solid foundational algebra skills for their future careers.

Evidence of what students actually know and have little or no conceptual understanding of has been reported in many research studies over the years. The word "understanding" is used in the context of mathematics with two different meanings: *relational* understanding, knowing both what to do and why, and *instrumental* understanding, knowing a rule and being able to use it (Skemp, 1987). He referred to instrumental learning as "rules without reason." It is these alternative meanings of understanding that are at the root of many of the differing perspectives on teaching and learning. Smith (1996) argues that an instrumental approach to teaching mathematics provides a teacher with a robust sense of efficacy.

Many preservice teachers believe this is the only approach to teaching mathematics that will provide them with a sense of competence, proficiency, and know-how. The belief that if a student demonstrates skills proficiency and gets a correct answer, he or she "understands" the concept is not necessarily true.

Autobiographical descriptions of community college developmental algebra students' and preservice teachers' prior instructional experiences reveal the very different attitudes and beliefs about mathematics held by many students compared with that of their instructors and which prove very resistant to change:

> All throughout school, we have been taught that mathematics is simply plugging numbers into a learned equation. The teacher would just show us the equation dealing with what we were studying and we would complete the equation given different numbers because we were shown how to do it.
>
> My previous mathematics experience involved a teacher lecturing. Finding a formula to solve a problem was, in reality, the answer to the problem.
>
> I was taught "how" but not "why". I thought of math as a series of formulas, each of which should be followed in order to find an answer.
>
> When I took Calculus the differentials were what killed me. There were so many equations, and the teacher would go over how to get from one to the other. I tried to go with what I knew, and memorize them. However, that didn't work.

Many prospective teachers have a deeply ingrained procedural orientation to mathematics with its focus on "getting the correct answer" that they have learned to value above all. Changing this orientation is made more difficult when instructors fail to take into account students' prior knowledge and underestimate what their students are capable of. A challenge facing mathematics instructors at undergraduate institutions and community colleges is how to develop a sound conceptual grasp of foundational arithmetic and algebraic topics in these students within the context of their course content.

2.2 The Role of Prior Learning and Its Possible Problematic Consequences

One of the most important findings of cognitive science and brain research is that prior knowledge is the beginning of new knowledge. Ausubel, Novak, and Hanesian (1968) remind us, "The most important single factor influencing learning is what the learner already knows. Ascertain this and teach accordingly." Prior knowledge is a fact, and it is persistent. New experiences that build on prior experience are much better remembered, and what does not fit in prior experience is either not learned or learned temporarily and easily forgotten.

The notion of *met-before* (McGowen & Tall, 2010; Nogueira, De Lima & Tall, 2008; Tall, 2004) was introduced to focus on how new learning is affected by the learner's previous experiences and as a way of looking at the effects of prior learning that can support or impede new learning. Prior learning can be supportive in those instances where old ideas can be used to make sense in new contexts and problematic in contexts where the old understanding no longer works. Research studies have examined students' ability to modify their prior knowledge. Discontinuities encountered were reported not only at the developmental arithmetic and

Fig. 2.1 Miscommunication (Thompson, 1996)

algebra levels but also at the undergraduate college level (Davis & McGowen, 2007; McGowen, 1998; McGowen & Tall, 2013).

Prior knowledge of the minus symbol to indicate subtraction, while supportive for whole numbers and positive fractions, becomes problematic for many students when they encounter negative numbers. It becomes even more so in the context of algebra to indicate the additive inverse of an unknown ($-x$). The belief held by many students is that a minus sign in front of it indicates that it has a negative value—a belief that results in increasing difficulties in subsequent mathematics courses (Fig. 2.1).

Ideas encountered at one stage of learning may lead to ways of thinking that are not appropriate later. Mathematics instructors need to take into account the effects of existing knowledge—both positive and negative—that students have *now* as a result of experiences they have met before, *at every level of development*, aware of how earlier mathematical experiences result in students' ideas that can become problematic when context and/or subtle changes in meaning are encountered.

2.2.1 Prior Arithmetic Thinking

The problematic nature of prior arithmetic thinking was revealed when a majority of 128 college freshmen, given the numbers 0, 1, x, y, and $-z$ as marked on the number line below (Fig. 2.2), claimed that $2y$ was larger than y because (a) "$2y$ is larger than y" or (b) "$2y$ is larger because it has a number in front of the variable."

Nearly one-third of the students maintained that $x-y=y-x$, stating that they were "the same problem just switched around" or "because they are both subtracting a variable." One-fifth of the group wrote that $x+y=x-y$ "because adding a positive and a negative is the same as subtracting a positive and a negative number." The other most common response was "because in both equations you are really adding the numbers" (McGowen & Tall, 2010, pp. 175–176).

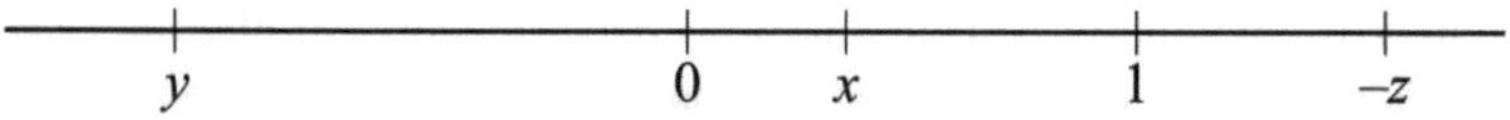

Fig. 2.2 Quantities on a number line (Bright & Joyner, 2003)

Students' efforts to interpret algebraic and function notation demonstrate how very differently individual students think about notation than do mathematicians. The conceptual requirements for understanding ambiguous expressions, both arithmetic and functional, appear to be far more formidable in their complexity than has generally been recognized by mathematics instructors. Sfard reminds us: "Algebraic symbols do not speak for themselves. What one actually sees in them depends on the requirements of the problem to which they are applied. Not less important, it depends on what one is *able* to perceive and *prepared* to notice" (Sfard, 1991, p. 17).

That students have difficulties with fractions, negative numbers, decimals, and the transition to algebra is well documented. How the changes in meaning of the successive number systems *N*, *F*, *Z*, *Q*, and *R* impact individual students at the undergraduate level—particularly where previous experiences involve some aspects that are supportive and generalize while others are problematic and impede new learning—has been less widely researched and reported. The multiple meanings of the minus symbol and the need to interpret symbolism flexibly were identified as major sources of difficulties for undergraduates, particularly when the symbolism changes meaning in new contexts (Davis & McGowen, 2007; McGowen, 1998; McGowen & Tall, 2013).

The minus symbol becomes problematic for many students when they encounter negative numbers, the precedence of division and multiplication over addition, and the precedence of powers over taking the additive inverse (McGowen, 1998). They are faced with new conventions when combining two operations: (a) the minus sign and the power operation, and (b)the order of operations of two *unary* processes, squaring negative three and taking the additive inverse of the square of three. Their prior knowledge consists of the order of *binary* operations and a mnemonic such as Please Excuse My Dear Aunt Sally (PEMDAS) to indicate the order of priority of operations (parentheses, exponent, multiply, divide, add, and subtract). Students who focus on the qualitatively different features of -3^2 and $(-3)^2$ are able to make sense of the notation and correctly manipulate these expressions. These students connect new knowledge with their prior knowledge in a way which results in a reconceptualization of the two processes of squaring a negative number and finding the opposite of a number squared. A student's typical explanation is:

> When I see the sign (−) it is a change for me to know that it means "the opposite of," I always though it meant a negative number or, $-(-x)$ *a positive* x. The reflection assignment enhanced my understanding of the opposite of a square by looking at it as two functions, and then order of operations would have exponents first, then the opposite of the value … Exponentiation takes precedence over oppositing in the absence of grouping symbols.

Data on students' difficulties evaluating -3^2 and $(-3)^2$ have been collected during various research studies. Pre- and post-surveys administered over several years[1] to 516 community college and university students enrolled in a developmental algebra course showed that, at the completion of the course, 81 % (418/516) of the students correctly evaluated $(-3)^2$, but only 49 % (251/516) of the students could correctly evaluate -3^2. The latter may simply relate to the way that the symbols are read from left to right "minus" "3" "squared." The minus and the three are taken together as "minus three," and the student has been told that the square of a negative number is positive.

In an expression with more terms to put together, the problem may become more complicated. When given $f(x) = x^2 - 3x + 5$, find $f(-3)$; a College Algebra student explained his work as follows:

> "*I used up the negative sign. I have to do parentheses first*" and wrote: $f(-3) = -3^2 - 3(-3) + 5 = 9 + 9 + 5 = 23$ followed by: $9 + 5$. "*Now I have to do this* (indicates the -3^2), "*but I can't remember if it's negative nine or just nine. I never know which to use.*" He wrote down -9 and stopped. "*There's no sign in front of this* (pointing at $9 + 5$), *so I need to multiply*," writing $-9(14) = 136$.

2.2.2 Interpreting the Minus Sign in Linear Factors

Many students are confused as to whether the minus sign in $(x - c)$ represents subtraction or is the sign attached to c. An instructor who participated in a recent formative assessment pilot study reported that many of her College Algebra students had difficulty using the linear factors of a polynomial correctly (McGowen & Tall, 2013). Working from a graph using the zeros of a function to determine a quadratic function's linear factors, several students viewed c in $(x - \mathrm{c})$ as a negative value, a belief which results in many sign errors when writing the factors and/or zeros of a function (McGowen, 1998):

> The value of c is negative because of the minus sign in front of c. c will subtract from any number that comes before the "−" symbol.
>
> I used up the negative sign.

Students used the subtraction operator of a linear factor as the sign of c, and many thought the x- and y-intercepts were the coefficients in the equation of the function. Given the graph of a line, only one in five students was able to determine whether the slope and y-intercept should be positive or negative. Inflexible in their thinking, they were only able to answer questions from one direction—unable to reverse the process.

Prior experience involving the minus sign also proves problematic for undergraduates in other contexts. Students are faced with making sense of function notation as well as interpreting the minus symbol in expressions such as $f(-x)$ and $-f(x)$. Many students believe that $f(-x)$ represents "f of negative x" or "a

[1] Portions of the collected data have been previously reported (McGowen, 1998; McGowen & Tall, 2013), but the accumulated data of 516 students have not been reported previously.

negative input value" and that $-f(x)$ represents "negative f of x" or "a negative output value." Some students interpret $-f(x)$ as "the entire function is negative" and $f(-x)$ as "only the x is negative." There are also students who interpret function notation as indicating multiplication: $-f(x)$ means $-f$ times x and $f(-x)$ as f times $-x$.

2.2.3 Understanding Basic Algebraic Terms and Concepts

A lack of understanding of basic algebraic terms contributes to students making other errors. College Algebra students' written comments indicate that many of them have only a vague understanding of the meanings of foundational concepts, such as slopes, coefficients, and intercepts. Given the graph of a line, only one in five was able to determine whether the slope and y-intercept should be positive or negative. Some interpreted *slope* as an ordered pair and plotted it as an intercept. Other students equated the x- and y-coefficients in the equation with the intercepts on the graph. Several students wrote that an intercept is a number value. Many of them wrote the slope as an ordered pair and did not view slope as a ratio. Still others used the value of the slope as the x-intercept value. They were only able to answer questions from one direction—unable to reverse the process—indicating the inflexibility of their thinking (Davis & McGowen, 2007).

In their capstone course, a class of senior mathematics majors intending to be secondary math teachers were asked to describe what they knew about slope, covariation, rate of change, tangent, and derivative and which, if any of these ideas, are related. Only one student identified *covariation* as the fundamental link among the five ideas. Many made no attempt to provide any meaning for covariation as it was not a recognizable term in the curriculum. Rate of change was connected to derivative only because derivative is an instant rate of change and only a few students explicitly mentioned what changed.

2.3 Identifying and Addressing Problematic Prior Met-Befores

Black and William (1998) examined approximately 250 studies and found that gains in student learning resulted from a variety of methods all of which had a common feature: formative assessment (assessment that uses the data acquired to adapt instruction to better meet student need). They found that when teachers understand what students know and how they think and then use that knowledge to make more effective instructional decisions, significant increases in student learning occur. For instructors at all levels unaware of the knowledge and understanding of basic mathematical concepts and terms students lack when they enter our courses, Krutetskii's advice is appropriate: "Don't make a hasty conclusion about the incapacity of children in mathematics on the basis of the fact that they are not successful in this subject. First, *clarify the reason for their lack of success*" (Krutetskii, 1969, p. 122).

Clarifying the reasons for a student's previous lack of success—identifying *what precisely is lacking* in an individual student's development—is a challenge facing mathematics instructors. If students' prior knowledge (met-befores) impedes new learning and has resulted in misconceptions, instructors need to adapt instructional strategies that overcome and transform students' problematic met-befores. Some classroom assessment strategies that have proven effective in identifying what students understand and useful in addressing problematic met-befores are:

- Explicitly discussing prior understandings and how it changes in a different context
- Asking basic questions that instructors assume students know the answers
- Comparing students' written responses to two or more questions dealing the same concept or with related concepts, revealing of their ability to think flexibly
- Pre- and posttesting that offers a measure of individual student growth over time
- Using a function machine representation and the graphing calculator to make sense of notation and a deeper understanding of binary and unary arithmetic operations
- Identifying one's own met-befores and examining how they impact one's teaching and beliefs about curriculum and students

As Thompson (1994) reminds us:

> An instructor who fails to understand how students are thinking about a situation will probably speak past their difficulties. Students need a different kind of remediation, a remediation that orients them to construct the situation in a mathematically more appropriate way.

2.3.1 Asking Basic Questions

Asking questions about basic mathematical concepts and terms of which one assumes students have good understanding often reveals problematic knowledge which interferes with new learning. A lack of understanding of basic mathematical terms like "solve" and "evaluate" is often not recognized. Many students believe that they "solve an equation" whenever "x" is part of an expression. As part of a formative assessment pilot project at a local community college, undergraduates were asked to complete the following:

A. Finding the output when the input is known is the process of:

(a) simplifying (b) evaluating (c) factoring (d) solving
B. Finding the input when the output is known is the process of:

(a) simplifying (b) evaluating (c) factoring (d) solving

Only nine of 75 (13 %) Introductory Algebra students and 15 of 114 (12 %) College Algebra students selected (b) *evaluating* as the correct response to question A. On question B, less than 19 % of Introductory Algebra students and only 31 % of College Algebra students chose option (d) *solving*.

Developmental algebra students at a community college participating in a recent online survey were asked, “What does it mean to solve an equation?” The response “to get a variable by itself” was given by 35 % of the participating students. A second question asked “Is $x=y$ an equation: Why or why not?” The most common student responses were as follows: “No, because you need a number”; “No, because there are no known numbers on both sides”; and “No, because x and y are variables.”

It is not only developmental algebra students enrolled in precollege courses that lack understanding of basic mathematical concepts and processes. A Ph.D. student completing his doctorate in mathematics and working as an online tutor for a textbook company when asked to explain the difference between solving an equation and evaluating an expression replied:

> If a book asks you to evaluate $x^2 - 2x + 1$, what they are asking for is a simplified version of this polynomial, which would be $(x-1)^2$.
>
> Solving an equation or expression is actually plugging in a particular value to come up with a solution.
>
> For example:
>
> $$F(x) = x^2 - 2x + 1 \text{ Solve for } f(4).$$
>
> $$F(4) = 4^2 - 2(4) + 1 = 16 - 8 + 1 = 9$$
>
> Is this helping you feel a little bit better about the difference between the two? (McGowen, 2006, p. 25)

2.3.2 Comparing Responses to Two or More Questions

The inability of students to correctly answer *two or more* questions on related content suggests that students do not see the questions as being intimately connected. One indication that what students have learned and remembered is fragmented and unconnected is that they are unable to apply what they know when *confronted with a different context.* Noticeable differences in students’ responses to related questions dealing with slopes of linear equations were reported in a study by Davis and McGowen (2007).

The questions and responses of 92 community college students enrolled in an Introductory Algebra course are shown below (Table 2.1). Note that only 20 of the 92 students were able to answer three of the five questions correctly, suggesting these students lack robust understanding of the mathematics they are learning (p. 24).

Table 2.1 Related responses to questions on linear equations, slopes, and intercepts

Question	$n = 92$	Correct (%)
7. Given slope −3 and y-intercept 5, select linear equation	65	71
4. Determine the x-intercept of the equation $2x - 7y = 12$	38	41
5. What is the vertical intercept of $y = mx + b$?	37	40
3. What is the slope of $Ax + By = C$?	24	26
9. Given the view window and graph, what is the equation?	13	14
Three of five questions answered correctly	20	22
Four of five questions answered correctly	6	7
All five questions answered correctly	1	1

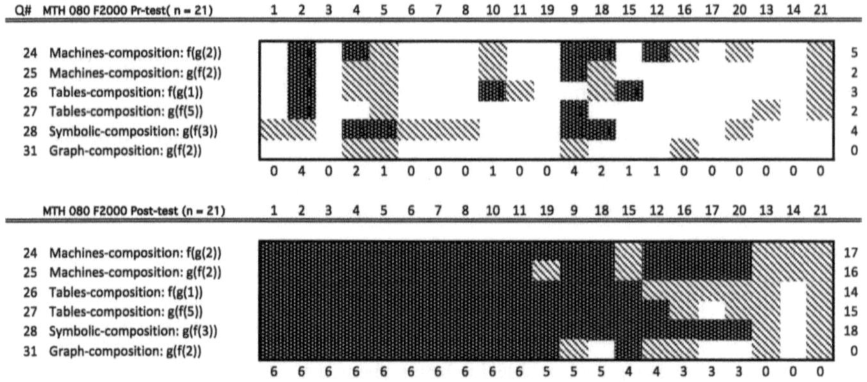

Fig. 2.3 Pre- and posttest responses on various representations of function composition

2.3.3 Comparing Pre- and Posttest Responses

A visual analysis comparing individual students' pre- and posttest responses to related questions is informative for students as well as for their instructor. Shown in Fig. 2.3 is an example of an analysis of students' responses to questions on various representations of function composition. An analysis of all pretest responses is shared with each student, identified by the column number which corresponds with a number on the individual student's returned test. Each student receives an analysis comparing all pretest responses to all posttest responses near the end of the course. Each column represents an individual student's responses, and each row represents the responses to a given question. A black cell indicates a correct response, a striped cell indicates an incorrect response, and a blank cell indicates no attempt to answer the question.

2.4 Function as an Organizing Lens: A Function Machine Representation and Technology

A NSF-funded developmental algebra curriculum (DeMarois, McGowen & Whitkanack, 1996a) has been shown to deepen developmental algebra students' understanding of mathematics, make sense of mathematical notation that increase skills proficiency, and provide them with opportunities to examine and reconstruct problematic prior learning (Davis & McGowen, 2002; DeMarois, 1998; DeMarois & McGowen, 1996b; McGowen, 1998; McGowen, DeMarois, & Tall, 2000; Tall, McGowen, & DeMarois, 2000). The unifying concept of function and difference equations facilitated students' ability to see connections and link fundamental ideas. Constant finite differences and ratios were used to develop sequences as functions, determine parameters, and develop models of linear, exponential, and polynomial functions.

A significant finding from this research on Introductory and Intermediate Algebra students is that initially there was a spectrum of interpretations from students who saw only a process, such as $2+4$ meaning "two is added to four," to those who could view notation flexibly, seeing the expression $2+4$ not only as a process but as the concept, "a sum," and $2x+5$ not only as the process of addition but also as the concept "expression." Students who successfully completed the course were found to be prepared for continued study of increasingly sophisticated mathematical ideas in both STEM and non-STEM courses.

Using the function concept and the graphing calculator results in a coherent sequence different from the traditional ordering of algebraic topics as shown below in Fig. 2.4. Developmental algebra students were able to make connections and generalize algebraic linear, exponential, and quadratic models from data. They gained a deeper understanding of parameters and increased flexibility of thinking.

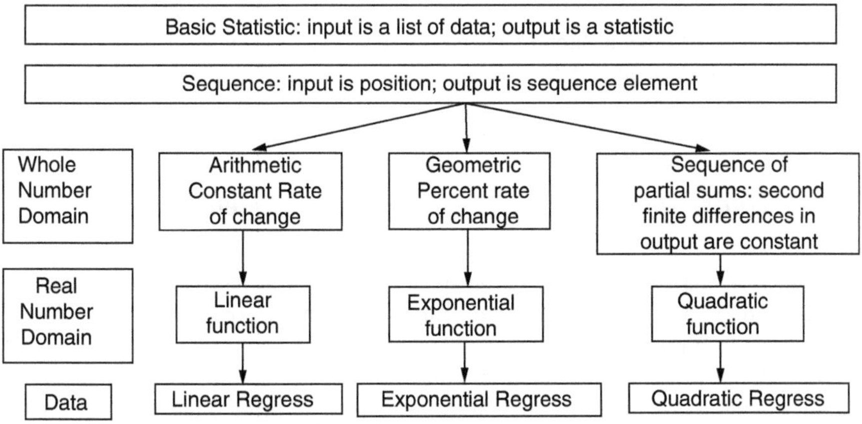

Fig. 2.4 Intermediate algebra course sequence

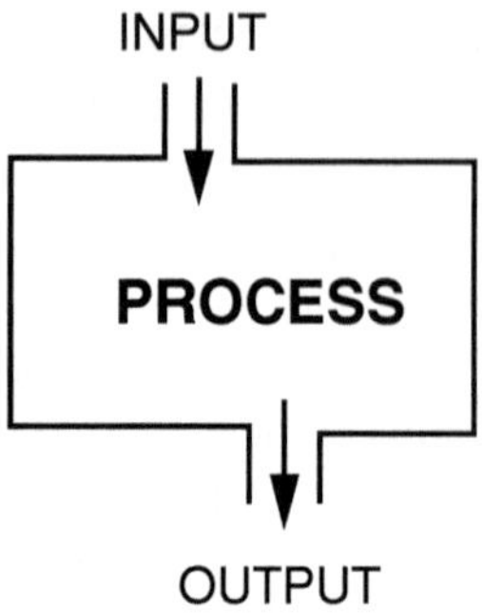

Fig. 2.5 Function machine representation

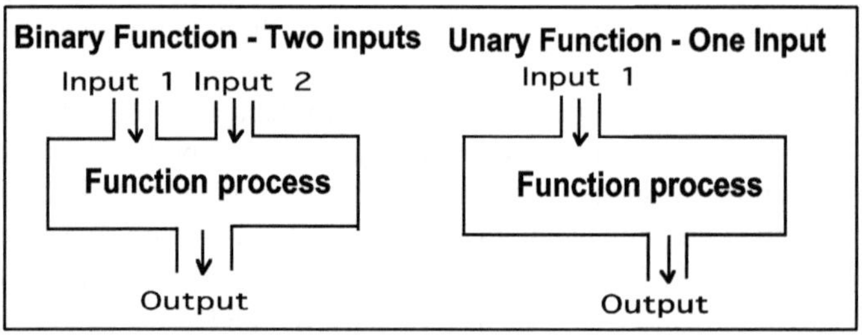

Fig. 2.6 Function machine representations: binary and unary processes

Functions viewed as input-output machines were studied in mathematics education as far back as 1965 by Peter Braunfeld. The function machine representation shown in Fig. 2.5, introduced as a visual representation for the concept of function seen as an input/output process, was found to be an accessible starting point for many developmental students (Davis & McGowen, 2002; DeMarois, 1998; McGowen, 1998; McGowen, 2006; McGowen, DeMarois & Tall, 2000). The function machine representation becomes a meaningful unit of core knowledge leading to more meaningful understanding of function, domain, range, and notation and was found to be a supportive met-before (McGowen & Tall, 2010; Tall, McGowen & DeMarois, 2000).

2.4.1 Understanding Binary and Unary Arithmetic Processes

The function machine representation in Fig. 2.6 is an effective visual means of distinguishing between binary and unary arithmetic operations (DeMarois, McGowen, & Whitkanack, 1996a, p. 23).

Students also find the function machine representation, together with the graphing calculator, helpful when reexamining their prior understanding of the minus sign and how meaning changes in different contexts. In their text, DeMarois,

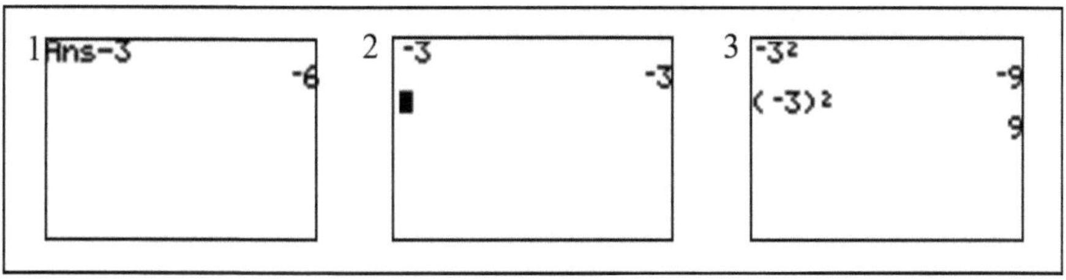

Fig. 2.7 TI-83 view screen of binary and unary operations

McGowen, and Whitkanack (1996a) discuss three distinct meanings of the minus sign: a binary arithmetic function requiring two inputs (subtraction), an object (a negative number), and a unary function requiring a single input (the "opposite of x" denoted by $-x$). Two different calculator keys input the minus sign: one indicating a binary operation, subtraction, and the other signifying a negative number or a unary operation, the additive inverse. Together with the function machine representation, students investigate three tasks: (1) subtract three; (2) type in the opposite of three; and (3) compare the results of entering -3^2 without using brackets and with brackets, typing $(3)^2$ (Fig. 2.7).

Pre- and posttests which ask students to evaluate $(5)^2$ and -5^2 were administered to two cohorts of students: 121 community college Intermediate Algebra students using the function machine approach and 140 university students using a traditional textbook. The responses of both cohorts were reported by McGowen and Tall (2013). Overall, both sets of students perform poorly in solving both questions correctly on the pretest. Both groups improved from pretest to posttest, with the community college students who had experienced the function machine strategy and the graphic calculator, improving more.

2.4.2 *Clarifying Understanding of Terms*

Students' understandings of "evaluating an expression" and "solving an equation" are clarified by examining them as functional processes, shown in Fig. 2.8.

In a preservice elementary teacher content course, an initial focus on building connections between different representations of a problem of binary choice led to students describing connections between building towers, grid walks, and binomial expansions. Followed by investigations of sequences as functions similar to those in the developmental algebra course, the prospective elementary teachers demonstrated deeper understanding of content, improved ability to generalize various arithmetic and geometric sequences, and improved skill competency, as well as changed attitudes and beliefs about mathematics (Davis & McGowen, 2002).

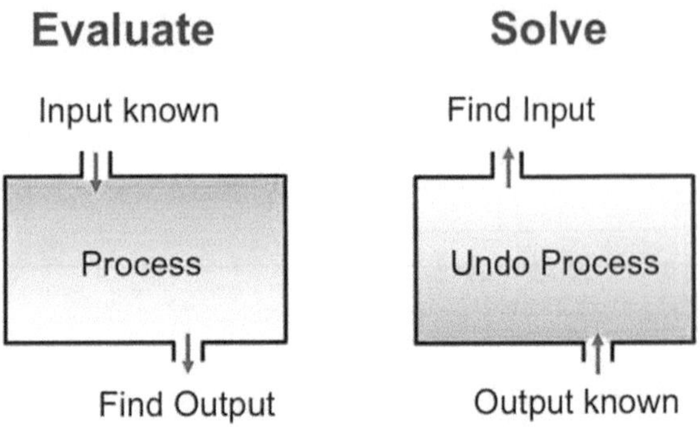

Fig. 2.8 Comparing the processes of evaluating vs. solving

2.5 An Open Question: Which Computational Skills Are Essential with Today's Technology?

The effective utilization of graphing calculators in the teaching and learning of mathematics has not yet been incorporated into the classroom by many instructors of developmental mathematics. Which pencil and paper skills students in a given course must be able to demonstrate remains an open and divisive question in mathematics departments, given the technology available today. Instructors, depending upon their beliefs and assumptions, have different orientations about the purpose and use of graphing calculators. Some instructors consider graphing calculators only useful as a means to check homework and do not permit their use on exams, fearing students' computational skills will deteriorate. Some use them to teach traditional mathematics topics sequenced in the traditional order with a more efficient, dynamic, or appealing presentation. Still others realize that using the graphing calculator effectively transforms the curriculum, necessarily altering the character of knowledge as well as the sequence and content of the curriculum, thus raising questions as to which computational skills are essential.

Heated exchanges still occur among instructors who hold differing beliefs about what students should know. At a university faculty workshop, participants were provided with the work of a developmental algebra student on the following problem and asked how they would evaluate the student's work:

> A toy rocket is projected into the air at an angle. After 6 seconds, the rocket is 87 feet high. After 10 seconds, the rocket is 123 feet high. After one-half minute, the rocket is 63 feet high.
>
> (a) The model for the rocket's motion is $h(\text{t}) = at^2 + bt + c$ where h is the height in feet of the rocket after t seconds. Using the given information, find the values for a, b, and c, so the function models the situation. Briefly explain what you did.
> (b) Approximate how long it will take for the rocket to hit the ground. Explain how you arrived at your answer.

The student set up a linear system in three variables by creating an input/output list on his calculator and used quadratic regression to find the parameter values. He

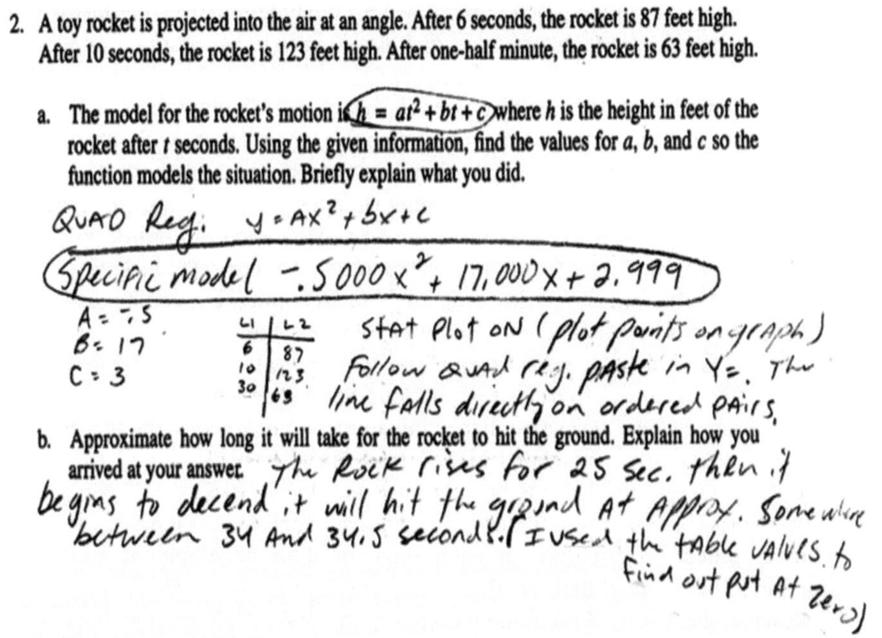

2. A toy rocket is projected into the air at an angle. After 6 seconds, the rocket is 87 feet high. After 10 seconds, the rocket is 123 feet high. After one-half minute, the rocket is 63 feet high.

a. The model for the rocket's motion is $h = at^2 + bt + c$ where h is the height in feet of the rocket after t seconds. Using the given information, find the values for a, b, and c so the function models the situation. Briefly explain what you did.

QUAD Reg: $y = Ax^2 + bx + c$

Specific model $-.5000x^2 + 17.000x + 2.999$

A = −.5
B = 17
C = 3

L1	L2
6	87
10	123
30	63

Stat Plot ON (plot points on graph) Follow quad reg. paste in Y=. The line falls directly on ordered pairs.

b. Approximate how long it will take for the rocket to hit the ground. Explain how you arrived at your answer. The Rock rises for 25 sec. then it begins to decend it will hit the ground At Approx. Somewhere between 34 And 34.5 seconds. (I used the table values to find out put at zero)

Fig. 2.9 Intermediate algebra student's work

then wrote the equation that models the situation, describing how far the rocket would rise before it began its descent and when it would hit the ground. His work is shown below (Fig. 2.9).

Faculty participants discussed how they would grade this work. Some instructors maintained that they would give the students no credit because he hadn't solved the linear system algebraically. Other instructors argued that the student should receive full credit as the response demonstrated very good understanding of the problem and his responses were correct. When the workshop facilitator asked participants: "Given the technology available today when will students be asked to solve a 3×3 linear system using pencil and paper outside of the classroom?" No one provided an answer to the question.

2.6 Conclusion

Teachers at all levels are faced with the result of the accumulated detritus of students' fragmented prior knowledge as a result of their earlier mathematical experiences. Though teachers cannot be expected to deal with all the problems that arise from previous learning, *at each stage* they need to be aware of the problematic met-befores of their individual students and counsel them accordingly. The failure to understand how students are thinking results in speaking past their difficulties and increasing disaffection on the part of students. Effective instruction is dependent on *how students use what they have learned before* and how what is taught *will affect what the students learn later*.

The problematic met-befores discussed in this chapter, including prior learning of arithmetic and the ambiguous minus symbol when faced with changes in meaning in algebraic and function notation, illustrate some of the complex aspects of learning algebra. A lack of conceptual understanding of the meaning of "solving an equation" compared with "evaluating an expression" and the failure to distinguish slope from x- and y-intercepts or coefficients contribute to the lack of success. These problematic aspects of learning algebra generally go undiagnosed and unaddressed by instructors and curriculum developers at all levels.

Explicit rethinking and reflecting on the longer-term effects of learning are essential if we are to improve student learning and success at every level—from the elementary grades through college and university. A focus on the specific changes in meaning as mathematics becomes more sophisticated and contexts change the meaning of what has been learned previously is a critical component in these efforts. Incorporating the use of formative assessment is essential in order to clarify students' difficulties at each stage of their development and adapt instruction to better meet their needs.

Following the Curriculum Foundations report, several alternative innovative courses designed to better meet the needs of students in technical and non-STEM career programs have been developed and adopted in many undergraduate and community college programs. A similar initiative is required to address the critical mathematical needs of our many future teachers and non-STEM students who will be required to have a much more solid mathematical foundation in arithmetic and algebra than they graduate with currently. As Einstein is reported to have said:

> Insanity is doing the same thing over and over again and expecting different results.
>
> We can't solve problems by using the same kind of thinking we used when we created them.

Currently lacking for students who need a deeper understanding of arithmetic and algebra are developmental courses and tasks designed to change the procedural orientation and superficial, fragmented knowledge of too many of our students. The curriculum, particularly for non-STEM undergraduate students along with those who intend to become teachers of mathematics, whether at the elementary, high school, or college level, should include a course that requires students to examine the long-term cognitive development of mathematical thinking and understanding of foundational arithmetic and algebraic concepts. Such a course would provide experiences in which students identify and analyze prior experiences that support new learning and identify situations in which prior learning can become problematic or be supportive.

Changing students' beliefs about the nature of mathematics and what it means to learn mathematics remains as much of a challenge today as it was more than 500 years ago when Robert Record (1543), in *The Grounde of Artes*, described the danger of rote learning and the resulting conflicting perspectives of teachers and students:

> Master: . . .I wil propounde here ii examples to you whiche if you often doo practice, you shall be rype and perfect to subtract any other summe lightly. . .
>
> Scholar: Sir, I thanke you, but I thynke I might the better doo it, if you did showe me the workinge of it.
>
> Master: Yes, but you muste prove yourselfe to doo som thnges that you were never taught, or els you shall not be able to doo any more than you were taught, and were rather to learne by rote (as they cal it) than by reason.

Mathematicians and mathematics educators must be willing to adjust their beliefs and assumptions about students' learning and incorporate what is known about the long-term cognitive development of mathematical thinking into their instructional practices. As Stephen Crane (1972) wrote: "The wayfarer, perceiving the pathway to truth, was struck with astonishment. It was thickly grown with weeds. . . Later he saw that each weed was a singular knife. 'Well,' he mumbled at last, 'Doubtless there are other roads'."

Complaining about what students can't do is no longer an option. Too many students are failing. Our challenge is to provide tasks and opportunities for students to engage in explicit examination of their prior knowledge when confronted with new situations and contexts. Recognizing that learning "just algebra" is a much more complex task than it appears. Incorporating what is known about the long-term cognitive development of mathematical thinking into instruction can result in changing how and what is taught so that more students who need to learn algebra can do so meaningfully and effectively.

References

American Mathematical Association of Two-Year Colleges. (2004). *Beyond crossroads: Implementing mathematics standards in the first two years of college*. Memphis, TN: American Mathematical Association of Two Year Colleges.

Attewell, P., Lavin, D., Domina, T., & Levey, T. (2006). New evidence on college remediation. *Journal of Higher Education, 77*(5), 886–924.

Ausubel, D. P., Novak, J. D., & Hanesian, H. (1968). *Educational psychology: A cognitive view.*

Bailey, T. (2009). Challenge and opportunity: Rethinking the role and function of developmental education in community college. *New Directions for Community Colleges, 145*, 11–30.

Barker, W., Bressoud, D., Epp, S., Ganter, S., Haver, B., & Pollatsek, H. (2004). *Undergraduate programs and courses in the mathematical sciences: CUPM curriculum guide, 2004.* Washington, DC: Mathematical Association of America.

Bautsch, B. (2013). Hot topics in higher education: Reforming remedial education. In *National Conference of State Legislatures.* http://www.ncsl.org/documents/educ/REMEDIALEDUCATION_2013.pdf.

Beaton, A. E. (1996). *Mathematics achievement in the middle school years. IEA's third international mathematics and science study (TIMSS).* Chestnut Hill, MA: Boston College.

Black, P., & Wiliam, D. (1998). Assessment and classroom learning. *Assessment in Education, 5* (1), 7–74.

Blair, R. M., Kirkman, E. E., Maxwell, J. W., & American Mathematical Society. (2013). *Statistical abstract of undergraduate programs in the mathematical sciences in the United States: Fall 2010 CBMS survey* (pp. 131–136, Table TYE4).

Breneman, D. W., & Haarlow, W. N. (1998). Remediation in higher education. A symposium featuring "Remedial education: Costs and consequences". *Fordham Report, 2*(9), 9.

Bright, G. W., & Joyner, J. M. (2003). *Dynamic classroom assessment: Linking mathematical understanding to instruction*. Vernon Hills, IL: ETA/Cuisenaire.

Brothen, T., & Wambach, C. A. (2004). Refocusing developmental education. *Journal of Developmental Education, 2*, 16–18, 20, 22, 33.

Carlson, M. P. (1998). A cross-sectional investigation of the development of the function concept. In A. H. Schoenfeld, J. Kaput, & E. Dubinsky (Eds.), *CBMS issues in mathematics education: Research in collegiate mathematics education III* (Vol. 7, pp. 114–162).

Cohen, D. (1995). *Crossroads in mathematics: Standards for introductory college mathematics before calculus*. Memphis, TN: AMATYC.

Common Core State Standards Initiative. (2010). *Common core state standards for mathematics*. Washington, DC: National Governors Association Center for Best Practices and the Council of Chief State School Officers.

Conference Board of the Mathematical Sciences. (2001). *The mathematical education of teachers. Part I and Part II*. Washington, DC: The Mathematical Association of America in cooperation with American Mathematical Society.

Crane, S. (1972). The wayfarer. In *The complete poems of Stephen crane*. Ithaca, NY: Cornell University Press.

Crane, S. (1972). The complete poems of Stephen Crane (Vol. 130). Cornell University Press.

Davis, G. E., & McGowen, M. A. (2002). Function machines & flexible algebraic thought. In *Proceedings of the 26th international group for the psychology of mathematics education* (Vol. 2, pp. 273–280). Norwick, UK: University of East Anglia.

Davis, G. E., & McGowen, M. A. (2007). Formative feedback and the mindful teaching of mathematics. *Australian Senior Mathematics Journal, 21*(1), 19.

De Lima, R. N., & Tall, D. (2008). Procedural embodiment and magic in linear equations. *Educational Studies in Mathematics, 67*(1), 3–18.

DeMarois, P. (1998). *Facets and layers of function for college students in beginning algebra* (Doctoral dissertation, University of Warwick).

DeMarois, P., & McGowen, M. (1996b). Understanding of function notation by college students in a reform developmental algebra curriculum. In *Proceedings of the 18th annual meeting of the North American chapter of the international group for the psychology of mathematics education, Panama City, Florida* (Vol. 1, pp. 183–186).

DeMarois, P., McGowen, M., & Whitkanack, D. (1996). *Applying algebraic thinking to data* (Preliminaryth ed.). Glenview, IL: Harper Collins Publishers.

Ganter, S., & Barker, W. (2004). The curriculum foundations project: Voices of the partner disciplines. *AMC, 10*, 12.

Krutetskii, V. A. (1969). Mathematical aptitudes. In J. Kilpatrick & I. Wirszup (Eds.), *Soviet studies in the psychology of learning and teaching mathematics* (Vol. II, pp. 113–128). Chicago, IL: University of Chicago Press.

McGowen, M. A. (1998). *Cognitive units, concept images, and cognitive collages: An examination of the processes of knowledge construction*. (Doctoral dissertation, University of Warwick). ERIC: ED466377. http://www.dissertation.com/book.php?method=ISBN&book=1612337732.

McGowen, M. A. (2006). Developmental algebra. In N. Baxter-Hastings (Ed.), *MAA Notes 69: A fresh start for collegiate mathematics* (pp. 369–375).

McGowen, M., DeMarois, P., & Tall, D. (2000). Using the function machine as a cognitive root for building a rich concept image of the function concept. In *Proceedings of the 22nd annual meeting of the North American chapter of the international group for the psychology of mathematics, Tucson, AZ* (pp. 247–254).

McGowen, M. A., & Tall, D. O. (2010). Metaphor or met-before? The effects of previous experience on practice and theory of learning mathematics. *The Journal of Mathematical Behavior, 29*(3), 169–179.

McGowen, M. A., & Tall, D. O. (2013). Flexible thinking and met-befores: Impact on learning mathematics. *The Journal of Mathematical Behavior, 32*(3), 527–537.

National Center on Education and the Economy (2013). What does it mean to be college and work ready? The Mathematics Required of First-Year Cmmunity College Students. Washington, DC. Available online at: http://www.ncee.org/college-and-work-ready.

National Council of Teachers of Mathematics. (1989). *Curriculum and evaluation standards for school mathematics*. Reston, VA: NCTM.

National Council of Teachers of Mathematics. (1991). *Professional standards for teaching mathematics*. Reston, VA: NCTM.

National Council of Teachers of Mathematics. (2000). *Principles and standards for school mathematics*. Reston, VA: NCTM.

National Research Council. (2000). Mathematics education in the middle grades: Teaching to meet the needs of middle grades learners and to maintain high expectations. In *Proceedings of national convocation and action conferences/center for science, mathematics, and engineering education*. Washington, DC: National Academy Press.

National Research Council. (2001). *Educating teachers of science, mathematics, and technology: New practices for the new millennium*. Washington, DC: National Academy Press.

Oehrtman, M. C., Carlson, M. P., & Thompson, P. W. (2008). Foundational reasoning abilities that promote coherence in students' understandings of function. In M. P. Carlson & C. Rasmussen (Eds.), *Making the connection: Research and practice in undergraduate mathematics* (pp. 27–42). Washington, DC: Mathematical Association of America.

Recorde, R. (1543). *The grounde of artes*. London: Reynold Wolff. http://www.alibris.com/The-Grounde-of-Artes-Robert-Recorde/book/14453565.

Sfard, A. (1991). On the dual nature of mathematical conceptions: Reflections on processes and objects as different sides of the same coin. *Educational Studies in Mathematics, 22*(1), 1–36.

Skemp, R. (1987). *The psychology of learning mathematics expanded* (Americanth ed.). Hillsdale, NJ: Lawrence Erlbaum & Associates, Publishers.

Smith, J. P., III. (1996). Efficacy and teaching mathematics by telling: A challenge for reform. *Journal for Research in Mathematics Education, 27*, 387–402.

Stigler, J. W., Givvin, K. B., & Thompson, B. J. (2010). What community college developmental mathematics students understand about mathematics. *MathAMATYC Educator, 1*(3), 4–16.

Stump, S. (1999). Secondary mathematics teachers' knowledge of slope. *Mathematics Education Research Journal, 2*(11), 124–144.

Tall, D. O. (2004). The three worlds of mathematics. *For the Learning of Mathematics, 23*(3), 29–33.

Tall, D., McGowen, M., & DeMarois, P. (2000). The function machine as a cognitive root for the function concept. In *Proceedings of 22nd annual meeting of the North American chapter of the international group for the psychology of mathematics education. Tucson, AZ* (pp. 255–261).

Thompson, P. W. (1994). Students, functions, and the undergraduate curriculum. In E. Dubinsky, A. Schoenfeld, & J. Kaput (Eds.), *Research in collegiate mathematics education. I. CBMS issues in mathematics education* (Vol. 4, pp. 21–44).

Thompson, P. (1996). If you say it, will they hear? In *A slide presentation given at the American Mathematical Association of two-year colleges annual conference, Long Beach, CA*.

Tucker, A. (1995). *Models that work: Case studies in effective undergraduate mathematics programs*. MAA Notes No. 38.

Part II
College Algebra in a Broader Context

Chapter 3
Long-Term Effects of Sense Making and Anxiety in Algebra

David Tall

Abstract This chapter offers a framework for the long-term development of sense making and anxiety for mathematics in general and algebra in particular. While many may see the development of algebra building from the basic ideas of arithmetic and generalizing to algebraic techniques for formulating and solving problems, over the long-term increasingly subtle changes of meaning may give pleasure to some yet become problematic for others. The symbol "-2" starts off as an operation "take away 2" but later represents the concept of a negative number, "minus 2." The algebraic symbol "$-x$" however only represents a negative number if x is positive, and takes on the new meaning as the "additive inverse" of x. While some students find algebra a source of pleasure and delight as it grows in sophistication, others find it problematic and seek to rote-learn techniques in ways that lack meaning in more sophisticated contexts. Here we consider how successive experiences that individuals encounter effect long-term learning. Sometimes experiences that are supportive in one context may become problematic, leading to negative emotional reactions. The chapter considers how various visual and symbolic approaches involve specific supportive and problematic aspects. Sometimes curriculum design that reduces the level of difficulty can give short-term success yet inhibit long-term sense making. On the other hand, by reflecting on profound underlying structures ("crystalline concepts"), mathematical ideas may be constructed and connected in ways that offer long-term flexibility.

Keywords Sense making • Anxiety • Supportive met-befores • Problematic met-befores • Level reduction • Crystalline concepts

For some students, algebra involves pleasurable activities in seeing patterns and building the foundations for more advanced mathematics. For others it is a source of anxiety with little use in everyday life. This chapter offers a "joined up" framework for mathematics in general, and algebra in particular, to help understand the reality

---

D. Tall (✉)
University of Warwick, Coventry CV4 7AL, UK
e-mail: david.tall@warwick.ac.uk

S. Stewart (ed.), *And the Rest is Just Algebra*, DOI 10.1007/978-3-319-45053-7_3

of the full range of experience of the whole population, in particular, why some find algebra a pleasurable activity, while others find it a source of anxiety.

Personally I am committed to see algebra introduced in a meaningful and positive manner. However, this has yet to be achieved in the wider population despite many years of seeking ways of improving sense making in algebra all over the world.

Research has shown that, as individuals attempt to make sense of increasingly sophisticated ideas over the longer term, new ways of thinking are required that may be enlightening for some and problematic for others. These have a long-term cumulative effect that widens the gap between those who are successful and enjoy the mathematics, those who are determined to cope with new ideas that they may not fully understand, and those who become increasingly disaffected. While curriculum designers may focus on the positive sequence of achievement that is desired, the implementation of the curriculum depends not only on the teacher's understanding of mathematics but also on how individual learners attempt to make sense of the new ideas and how teachers can mentor them to develop more powerful understanding.

3.1 Long-Term Sense Making

While mathematics may be seen to be coherent from an expert viewpoint, this is not true over the longer term for most learners, nor is it true at the boundaries of mathematical research where mathematicians are grappling with new ideas. It is a salutary fact that all mathematicians enter the world as newborn children with brains as yet not sufficiently formed to make subtle connections. So everyone goes through a long-term process of making sense of increasingly subtle ideas.

At each stage, new mathematical ideas often require more sophisticated ways of thinking. In considering how we may make algebra more meaningful, it is not sufficient just to look at the status quo. It is essential to see how different individuals make sense of mathematics over the longer term. What is learnt at earlier stages and how it is interpreted by learners at each successive stage continually build up a broader spectrum of different ways of working that affect the attitudes and understandings of new ideas as they are encountered.

For example, in early arithmetic, the process of addition becomes more compressed, starting from three counting processes (count one set, count the other, put them together and "count all") to a single counting process ("count on" the second number after the first), then to a known fact that can then be used as part of a flexible knowledge structure to derive new facts from known facts. This produces a long-term bifurcation in performance as some students continue to use less sophisticated counting procedures that become more complicated, while others benefit from the flexible use of more sophisticated ways of thinking that are more productive (Gray & Tall, 1994).

This bifurcation continues as whole number arithmetic develops through more sophisticated topics: making sense of place value, multi-digit addition, subtraction, multiplication, long division of whole numbers, introducing fractions and decimals, negative numbers, and so on. As number systems become more sophisticated, the generalized properties of the operations are seen first as generalized arithmetic and then as manipulation of variables in algebra, combining operations with symbols, and visualization using graphs.

It is inevitable that long-term learning involves successive encounters with new ideas that behave differently from previous experience. This involves not just what the teacher intends to teach but also what the learner senses in the ideas encountered. These ideas may not be explicitly taught, but they may have a profound effect in the learner's sense of security in handling new experiences. For example, taking away a whole number always gives a smaller result, but this is not always true for signed numbers. Multiplication of whole numbers gives a bigger result, but this doesn't always happen with fractions. In each case, the operations become more complicated and may cause uneasy feelings as the nature of number is generalized. How can taking something away give more? How can "two minuses make a plus?" The square of a nonzero number is always positive, so how can there be a number i such that $i^2 = -1$?

When mathematics is extended to new situations, research tells us that students often have "misconceptions." The literature is so vast that it would be invidious to give a single reference. However, the same data may be analyzed in a new way to find that the "misconceptions" often involve using methods that worked perfectly well at one stage, yet, without reconstruction, fail in a more sophisticated situation.

For instance, the difference between two single-digit numbers effectively means "take the smaller one from the larger," but when a child meets two-digit subtraction written in columns, taking the smaller from the larger in each column may lead to an erroneous answer such as concluding that $43 - 27$ is 24 because the difference between 4 and 2 is 2 and the difference between 3 and 7 is 4. In such a situation, it seems evident that the learner needs to be taught to use the correct procedure, but is this sufficient to deal with successively more sophisticated procedures? Simply being taught how to perform an appropriate procedure without understanding may lead to greater difficulties being encountered at a later stage, causing even greater confusion over the longer term.

Subtle changes in meaning occur throughout the mathematics curriculum. For example, in shifting from arithmetic to algebra, a sum such as $2 + 3$ always has an answer (in this case 5), but in algebra, an expression involving letters as variables such as $2 + 3x$ does not. Here it is possible that a particular learner who has not yet made sense of the meaning of algebraic notation may look at $2 + 3x$ and recognize the first part $2 + 3$ as an arithmetic operation that *can* be performed to give 5, but then the remaining x cannot be incorporated, so the learner leaves the answer as $5x$.

In attempting to help learners make sense of new ideas, it is important for the teacher, as mentor, to be aware of the possible effects of previous learning. For instance, the famous "students and professors problem" (Clement, Lochhead, & Monk, 1981)—in which the number of students is S, the number of professors is P, and there are 6 students for each professor—is often written as $6S = 1P$. This provoked a whole array of research papers to analyze what is happening and how to deal with it, when the main reason is there for all to see. Letters are often used as units, for example,

$$1\,\text{m} = 100\,\text{cm}$$

to represent 1 m is 100 cm. Interpreting S as "students" and P as "professors", it is evident that 6 students correspond to 1 professor, so this leads to

$$6S = 1P.$$

Often letters are used as objects to introduce students to manipulation of expressions, so $6a + 3b$ is 6 apples and 3 bananas, which allows an expression such as $6a + 3b + 2a$ to be simplified to 8 apples and 3 bananas or $8a + 3b$. The student now *seems* to be able to manipulate algebraic expressions in simple cases and may use this interpretation to have initial success in manipulating algebra. However, in the longer term, this is likely to sow problematic seeds that can grow into the student-professor problem.

This is a recurring phenomenon throughout the curriculum as each individual interprets new ideas in terms of previous experiences that are sometimes supportive, giving increasing mathematical power, and sometimes problematic, causing increasing difficulty.

3.2 Supportive and Problematic Met-Befores

As we build on our previous experience, we use ideas that are familiar to interpret new experiences. Having developed language to describe certain familiar circumstances, the same language is available to describe similar ideas in new contexts. This has led to a range of theories in which human thinking is expressed in terms of metaphor (e.g., Lakoff & Johnson, 1980; Sfard, 2008).

The theory of metaphor can be very helpful in gaining insight into puzzling situations in mathematics learning. However, it has an Achilles heel: it is often used to consider the problem from the sophisticated viewpoint of the expert, and this may be very different from the wide range of thinking of different learners.

As I sought to understand what was happening from the viewpoint of the learner who has yet to develop the sophistication of an expert, I played with the sound of the word "metaphor" and invented the new word "met-afore" using the old English word "afore" to refer to ideas that the learner had met before. Then I replaced "met-afore" by the new term "met-before" which distinguished "metaphor" and "met-before" in sound as "metAphor" and "metBefore." It can be formulated as follows:

> A met-before is a mental construct that an individual uses at a given time based on experiences they have met before. (Lima & Tall, 2008)

The first publication using the term met-before (Tall, 2004) focused mainly on met-befores that cause difficulty, giving the limited impression that they simply refer to misconceptions. However, it is important to balance the positive and negative aspects to give a balanced view of how we use previous experience in new situations. I defined a *supportive* met-before to be a previous experience that supports learning in a new situation and a *problematic* met-before as a previous experience that causes difficulties. It is essential to see the met-before operating in a specific new context, as a particular previous experience may be supportive in one context and problematic in another. For instance, "take away gives less" is supportive in whole numbers and fractions (without sign), but problematic in signed numbers and much later in handling infinite cardinal numbers. The way in which a learner copes with a met-before can cause very different emotional reactions and consequent differences in future progress.

This phenomenon is not just restricted only to students, it occurs in experts too, as can be seen throughout history when firmly held beliefs are challenged by new possibilities that prove difficult to grasp. Our language is littered with terminology that reveals these transitions, from *natural* numbers to introduce *negative* numbers, from *rational* numbers to introduce *irrational* numbers, from *real* numbers to *complex* numbers that have real and *imaginary* parts.

As we consider the long-term sense making of our students, we need to be aware that the same mechanisms operate at different levels in the minds of all of us, including teachers, curriculum designers, and expert mathematicians. Moreover, the fact that we have each had our own personal developments in different communities means that we may see the learning of mathematics from very different perspectives and the way in which one community interprets mathematics may be appropriate or entirely unsuitable for another.

Mathematicians who have reorganized their thinking to a powerful expert level may have ways of operating that are good for them yet prove to be problematic for learners, while educators and teachers who present mathematical ideas to learners at a given stage may be unaware of the later consequences of their teaching. It is therefore important to develop an overall picture of mathematical development so that those in different communities of practice can be sensitive to both the long-term goals of learning and using mathematics and also to the personal development of the individual.

3.3 Long-Term Development of Mathematical Thinking

As mathematics grows in sophistication, both corporately in history and individually in each one of us, changes of meaning occur to deal with new situations. To understand how individuals cope with such changes, it is necessary to consider:

- The increasing sophistication of mathematics
- The personal interpretations of the individual
- The long-term effects of sense making in increasingly sophisticated contexts

Long-term development of mathematical thinking begins with practical experiences that develop into more theoretical ways of reasoning. As learners meet new ideas, their personal interpretations are affected by a succession of previous supportive and problematic met-befores that subtly affect their thinking.

Problematic ideas in algebra may have their origins in early arithmetic and accumulate through successive experiences over the years. If the problematic aspects remain unresolved, the spectrum of difficulties may become so complicated that it may no longer be easy to resolve problems arising in a particular topic because their origins are so deeply embedded in the subconscious mind of the learner.

For example, McGowen and Tall (2010, 2013) reveal how the minus sign changes its meaning as the curriculum advances through the years, first as an operation of subtraction, then as a sign to denote a negative number such as -3, then as a sign to denote the additive inverse $-x$ which could have a positive value if x is negative. Combine this with another operation, such as squaring, then "-3 squared" may be "the square of -3" or "the additive inverse of 3 squared" which is resolved by an appropriate use of brackets, but may, in practice, be confused by students struggling with college algebra (see McGowen, Chap. 2, this volume.)

This general phenomenon occurs throughout learning. At any stage, a given group of learners will contain individuals at different stages of development, so what some may grasp easily will be difficult or even impossible for others. Over time, as new challenges are encountered in new contexts, the spectrum of possibilities may become more diverse, and the teacher's task in helping students make sense of new ideas becomes more complicated.

3.4 Visualizing and Symbolizing

Manipulating objects to "see" relationships is a powerful way of getting a sense of more general properties in arithmetic such as commutativity of addition and multiplication. However, some visual ways of symbolizing algebraic relationships may be supportive at one stage but become problematic in more sophisticated situations.

Tall (2013), Chap. 7, considers the case of the algebraic identity

$$a^2 - b^2 = (a - b)(a + b).$$

This can be represented visually as a large square side a taking away a smaller square side b to see the relationship perceptually (Fig. 3.1).

This picture of the difference between two squares can be used to visualize the meaning of the equation. However, implicitly, the values of a and b are positive. What happens if one or both of a and b are negative or if a is *less* than b?

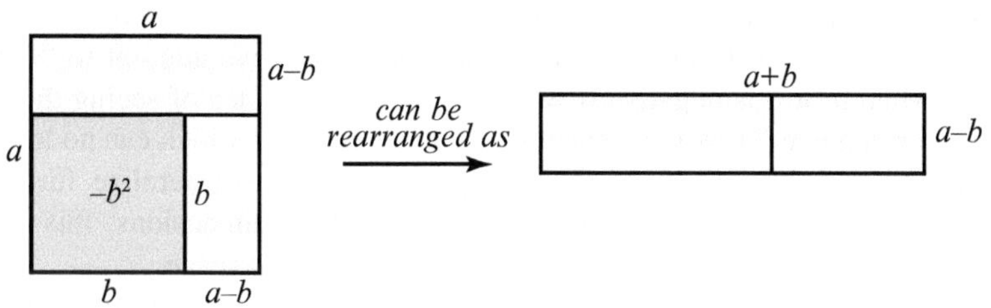

Fig. 3.1 Rearranging the difference between two squares $a^2 - b^2$

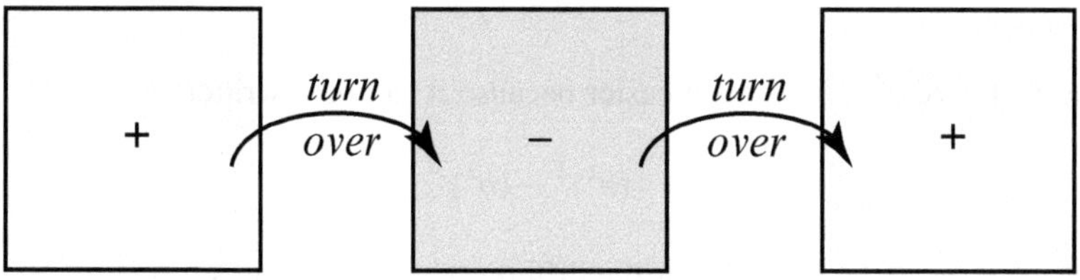

Fig. 3.2 Changing signs by turning over to see the reverse side

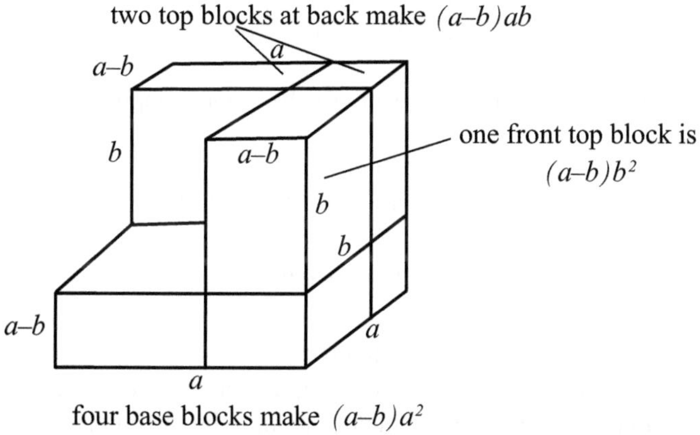

Fig. 3.3 $a^3 - b^3 = (a - b)\,ab + (a - b)\,b^2 + (a - b)\,a^2 = (a - b)(ab + b^2 + a^2)$

In this case, it is possible to "see" a change in sign as "turning over" the square to see the other side, interpreting one side as positive and the other as negative. We can now see that the operation of turning it over changes the sign, then turning it again returns to the original sign, so "two minuses make a plus" (Fig. 3.2).

When we shift from the difference of two squares in two dimensions to the difference between two cubes $a^3 - b^3$ in three dimensions, we can still see what happens when a smaller cube b^3 is removed from a larger cube a^3 (for a and b both positive) (Fig. 3.3).

This allows us to visualize the formula $a^3 - b^3 = (a - b)\ (ab + b^2 + a^2)$ which can also be found easily using symbolic algebra. But if we attempt to "see" the formula when a or b is negative or when $b > a$, then the idea of seeing the minus sign as "turning over" involves reflecting a cube in a mirror which can no longer be performed by a physical movement in space. When we generalize further the difference between two fourth powers $a^4 - b^4$ in four dimensions, this is even more problematic for creatures living in three-dimensional space.

As we generalize, the picture becomes more difficult to see. Yet if we focus on factorizing algebraically, it is possible to factorize

$$a^3 - b^3 = (a - b)(a^2 + b^2),$$

and to factorize $a^4 - b^4$ is even easier because it can be rewritten as

$$(a^2)^2 - (b^2)^2$$

and we can reuse the formula for the difference of two squares to get

$$\begin{aligned} a^4 - b^4 &= (a^2)^2 - (b^2)^2 \\ &= (a^2 - b^2)(a^2 + b^2) \\ &= (a - b)(a + b)(a^2 + b^2). \end{aligned}$$

So now, symbolic operations seem to be more appropriate than visual representations.

However, the advantage of symbolization over visualization is short-lived. When we consider $a^5 - b^5$, the factorization turns out to be

$$a^5 - b^5 = (a - b)\left(a^2 - 2\cos(72^\circ)ab + b^2\right)\left(a^2 - 2\cos(144^\circ)ab + b^2\right).$$

This factorization is unlikely to be found by manipulating algebraic symbols.

At a much later stage, when we have complex numbers at our disposal, we can "see" the complex roots of $z^n = 1$ which turn out to be the complex roots of unity of the form $e^{2\pi i/n}$. In the case of $n = 5$, this is where the values cos(72°) and cos (144°) arise as the values of 2π/5 and 4π/5 expressed in degrees (Tall, 2013, pp. 168–171).

The moral of this story is that in the long-term development of sophistication in mathematics, visual ideas can give insight, while symbolic ideas give increasing power that take us beyond our original perceptions. In later contexts, new kinds of visualization may give insights that shift us onto higher levels of operation.

3.5 Embodied, Symbolic, and Formal Development

So far, our analysis has been expressed in terms of visualization and symbolization, growing from practical activities in arithmetic where we can "see" more general relationships, then onto generalized arithmetic that can be symbolized using algebra and visualized using pictorial representations. However, these mathematical activities involve other forms of human perception and action than just visualization. They are based not only on our physical senses and actions but also our mental imagination. I use the term *conceptual embodiment* to include the full range of physical and mental conception and action (Tall, 2004). The inclusion of mental imagination in our embodied thought is essential to take us from our perception and actions in the actual world to more sophisticated mathematical concepts. For example, it includes the way in which we sense general properties of arithmetic operating on objects that form a basis for properties in algebra.

For instance, by acting on a set of, say, six objects physically or mentally, we may see that $4+2$, $2+4$, 2×3, 3×2 are precisely one and the same number. This allows us to sense that the results of addition and multiplication are unaffected by changing the order.

Embodied operations on objects such as counting, adding, subtracting, and so on may then be symbolized to develop a distinct form of mathematical thinking. Instead of imagining objects being moved around, the focus of attention switches to operating with the symbols themselves. This gives a new way of thinking that may be termed *operational symbolism.*

Here, the focus of attention changes from physical perception and operation on objects to operating with the symbols themselves. The symbols then may be conceived as mental objects that may be operated upon, and these new operations become mental entities that can be operated upon, and so on. Counting becomes number, addition of numbers becomes sum, repeated addition becomes multiplication, and generalized operations become algebraic expressions.

Initially, an equation of the form $3x+1=7$ may be seen as an operation "3 times a number plus one is 7." This may be "undone," first by taking off the 1 to get "3 times the number is 6" from which we can see that the number x must be 2. It only involves operations on *numbers*, taking 1 from 7 to get 6, then dividing by 3 to get the answer 2. It can be solved because there is a single algebraic operation "$3x+1$" which has a numerical result, and the solution can be found purely in terms of the operations of arithmetic.

However, an equation such as $3x+1=2x+3$ involves different operations on the two sides, and the arithmetical operation of "undoing" cannot be applied in such a simple way. On the other hand, we may embody this equation as a physical or mental "balance" where x is a quantity and 3 lots of x plus 1 balances 2 lots of x plus 3 (Fig. 3.4).

It is then possible to "do the same thing to both sides," first taking off 2 lots of x from both sides to get $x+1$ balancing 3, then taking off 1 from both sides to get $x=2$.

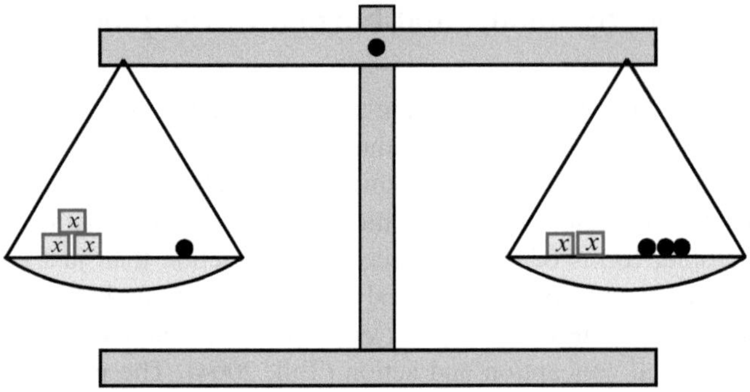

Fig. 3.4 Embodying the equation $3x + 1 = 2x + 3$ as a balance

This seems to support the idea that an embodied approach using a balance is more general than a symbolic approach because it can enable the learner to cope with more general equations. However this is not so because an equation such as $2x - 1 = 5$ cannot easily be represented as a physical balance as it has $2x - 1$ on the left, and we cannot take away the 1 from $2x$ as a physical move when we don't yet know what x is as a number.

As a consequence, we see that symbolic undoing and embodied balance are each supportive for some types of equation but problematic for others. Furthermore, there are some equations such as

$$3x - 1 = 2x + 1$$

which are not suitable for either symbolic undoing (it has expressions on both sides) or a balance model (it has a minus sign on the left).

A more sophisticated technique is the principle of "doing the same thing to both sides." First "add 1 to both sides" to get

$$3x = 2x + 2$$

then "take $2x$ from both sides" to get

$$x = 2.$$

A student who grasps this principle is likely to have an all-inclusive strategy to solve more general equations. However, it turns out that this more general approach is problematic for many students. When three classes of teenage students had been taught to find the solution by "doing the same thing to both sides," Lima (2007) found that none of them mentioned the general principle when interviewed at a later stage. Instead, many referred to the use of specific rules such as "change sides, change signs." The equation "$3x - 1 = 5 + x$" was solved by shifting the "-1" to the right-hand side to get "$3x = 5 + x + 1$," rearranging it to get "$3x = x + 6$," then after shifting the x to the right and simplifying to get "$2x = 6$," using a second rule:

"shift the number 2 over the other side and put it underneath" to get "$x = 6/2 = 3$." Common errors included misremembering the rules and mixing them up, such as solving $2x = 6$ by "moving the number underneath and changing its sign" leading to the error $x = 6/-3 = -2$.

This symbol shifting involving rules that may or may not be understood is termed "procedural embodiment." It is a procedure that is embodied by using the mental action of shifting symbols. If used correctly, procedural embodiment will lead to correct solutions. However, it can also break down and lead to error.

Procedural embodiment causes even more complicated difficulties at the next stage. For example, when the students in this study moved on to solve quadratic equations, the teachers, aware of the students' difficulties, focused mainly on giving them an apparently guaranteed solution by solving an equation in standard form $ax^2 + bx + c = 0$ using the formula

$$\frac{-b \pm \sqrt{b^2 - 4ac}}{2a}.$$

This failed because most of the students could not cope with the algebraic manipulation required to transform a quadratic equation into standard form (Tall, Lima, & Healy, 2014). The failure to make sense of symbol manipulation in linear equations has even more serious consequences in quadratic equations.

3.6 Level Reduction

We have seen how students encounter new situations that are either too complicated for them to make sense of the new ideas or where they learn a technique that works at one level but becomes problematic at the next. Concerned teachers, who wish to help their students reach attainment levels, may turn to routine practice of test items to encourage them to pass the next test. This may succeed in the short-term but in the long-term successive focus on procedural learning may prove increasingly problematic.

In his book *Structure and Insight*, Pierre Van Hiele (1986, p. 39) formulated the notion of successive levels of thinking:

> You can say someone has entered a higher level of thinking when a new order of thinking enables him, with regard to certain operations, to apply these operations to new objects. (translated from the Dutch original in Van Hiele, 1955, p. 289)

His work is usually associated with geometry where he introduced five levels, but in *Structure and Insight*, he was well aware that his broad framework of levels applies widely throughout mathematics and in other domains of knowledge. He declared that the number of levels was not important. What matters are the changes that occur in the shift in level from a familiar situation to a new level of thought. He settled on three basic levels: *visual*, *descriptive*, and *theoretical*. First situations are seen as a whole, then their properties are described, and then definitions are formulated in ways that can be used to build a theory deduced from those definitions.

This can be applied more generally by broadening the notion of visualization to incorporate the full range of embodiment using perception and action. In the development of arithmetic and algebra, the first basic level starts with practical arithmetic operating on physical objects in which we build the operations of arithmetic. In terms of the properties of arithmetic, we first *recognize* fundamental underlying properties, at the next level we *describe* those properties, and then we move to a *theoretical* level of definition of properties that are used to deduce a theory.

Practical examples may have limitations that are not found at the theoretical level. For example, temperatures above and below zero or as heights above and below sea level may be represented as signed numbers. In these practical examples, we can perform various operations. For example, we can shift up or down by a certain amount, but we do not actually add temperatures or elevations. We certainly do not multiply them. As a consequence, the theoretical level may involve new ideas that may not be represented in specific examples. On the other hand, the practical level may have differences that do not occur at the theoretical level. For example, in practice, three lots of two may be different from two lots of three (think, for example, of three ducks with two legs or two ducks with three legs). The shift from the practical to the theoretical involves various subtleties that may be problematic. But at the theoretical level, as pure numbers, 2×3 is precisely the same as 3×2. Hence if thinking operates at this higher level, the ideas are simpler than at the practical level. Students adhering to the specifics of practical examples therefore may find it problematic to shift to the theoretical level, while those who achieve the shift have a more flexible way of operating mathematically.

Van Hiele (1986, p. 53) used the term "level reduction" to describe how "it is possible to transform structures of the theoretical level with the help of a system of signs, by which they become visible" (Van Hiele, 1986, p. 53). By this statement, he refers to the possibility of manipulating symbols in a way that can be seen but not necessarily understood. For example, a student may learn to reproduce a geometric theorem by rote learning without grasping the underlying structure of the proof. In arithmetic or algebra, level reduction occurs when learning a procedure to carry out an operation or solve an equation without understanding what it means or why it works. A typical example is the use of procedural embodiment in solving equations.

The literature has long debated the distinction between different forms of knowledge and understanding, such as instrumental and relational understanding (Skemp, 1976) or between procedural and conceptual learning (Hiebert & Lefevre, 1986). Both types of learning and understanding play important roles in mathematics. Procedural learning is important to develop the necessary skills, while conceptual understanding provides a grasp of the bigger picture in formulating and solving problems.

If level reduction occurs at successive levels, the difficulties may be compounded. While learning procedures may give short-term success in passing the next test, the cumulative effect of several stages of level reduction may have more serious effects in subsequent learning. The short-term success focusing on procedural learning may lead to teachers and students changing their goals to the immediate priority to pass the test. In the longer term, it may become the default goal, with successive level reductions leading inevitably to a form of mathematics that is less suitable for more sophisticated thinking.

This is a phenomenon that can be seen around the world as comparison of success on various levels of testing causes the need to pay attention to achieving the highest possible grades at each level. This pressure can have an enormous detrimental effect on those who are not making sense of the mathematics at successive levels and so build a sequence of short-term rote learning to pass the test, with possible short-term success but with longer-term consequences.

3.7 Where Do We Go from Here?

Currently, governments around the world are concerned with maintaining and raising standards in mathematics as measured in international tests. This is leading to curriculum design formulated in terms of successive levels of subject development that are tested to judge the apparent level of achievement of children at various stages.

In the United States, the Executive Summary of the NCTM Standards (NCTM, 2004) presents a desired positive view of development through successive stages of learning. However, there is no mention of the problems that students face as they encounter new ideas in mathematics that may cause them difficulties, even in more recent versions (e.g., NCTM, 2013). The reason for this is that the summary is written to specify desired objectives, written by a group of experts working for the government (NGA, 2009), who are mainly involved in the testing industry (Schneider, 2014). The implementation of the standards is left to the professionals in mathematics education.

The NCTM (2013) issued a statement supporting the standards while acknowledging that they are "not sufficient to produce the mathematical achievement that our country needs to be competitive in the twenty-first century." To accommodate the standards within a wider framework, an additional list of positive initiatives were proposed to achieve that success. However, these initiatives again make no explicit mention of the problematic feelings toward mathematics that are said to affect around two thirds of the adult population (Burns, 1998). It is as if speaking only of the positive while remaining silent on the negative aspects that affect the majority of the population can lead to success.

History does not support this view. In recent times, there have been resounding calls to change everything for the good: the "new math" of the 1960s promised that if only we got the mathematics right, everything would be well. But the mathematics, as seen by experts, did not make sense to the wider population of learners. Then the constructivism of the 1980s turned attention to the learners to encourage them to make sense of mathematics in their own way, but this did not produce the desired results required in universities and in the workplace. In the new millennium—despite repeated calls to "raise standards"—the level of performance in the United States (and also in the United Kingdom and other Western countries) has failed to improve in international comparisons (Pisa, 2012).

Drawing together the lessons of the past may offer a way forward to enhance the future evolution of the teaching and learning of mathematics. While the "new math" focused on a modern approach to mathematics, the constructivist approach focused on the development of the individual. Now it is time to blend both together to develop an integrated approach by combining three aspects mentioned earlier:

- The increasing sophistication of mathematics
- The personal interpretations of the individual
- The long-term effects of sense making in increasingly sophisticated contexts

In the remainder of this chapter, we look successively at these three aspects. First, we consider the coherent structure of mathematical concepts that evolves as mathematics increases in sophistication in successive contexts by using the notion of "crystalline concept" (Tall, 2011). This offers a positive overall picture of development that will enable teachers to be aware of how their current teaching can help the student focus on long-term learning.

Second, we consider the personal interpretations of the individual where an awareness of supportive and problematic met-befores can assist the teacher and learner to focus on ideas that improve insight and guard against negative effects that impede long-term learning.

Finally, these two aspects will be blended together to suggest strategies for encouraging long-term sense making.

3.8 The Increasing Crystalline Sophistication of Mathematics

We have seen how arithmetic develops in sophistication and generalizes into the study of algebra. At each stage, operations are introduced and symbolized so that the symbols themselves can be manipulated as mental entities at a higher level. For instance, the process of counting leads to the concept of number, and then operations can be performed on number such as addition, subtraction, multiplication, and division to lead to new concepts of sum, difference, product, and, in the final case, sharing in the context of whole numbers and fractions in the context of dividing objects into smaller parts.

Gray and Tall (1994) noted that the same symbol could be used to represent both a process and a concept: $2+3$ as the process of addition and the concept of sum, 2×3 as multiplication and product, $\frac{2}{3}$ as sharing and fraction, and $2+3x$ as a general operation of adding 2 to 3 times x and as an expression which could itself be manipulated. At the same time, different processes could give rise to the same concept, for instance, $2+3$ is the same as $3+2$ or $1+4$ or $6-1$. The term "procept" was used to speak of the underlying entity that could be manipulated flexibly in many ways.

As the curriculum progresses, procedures that are seen initially as being different are seen as a rich blending of different processes giving the same procept. Counting a particular set that can be performed in different ways but always gives rise to the same number. A whole number is a procept that has a rich internal structure where a number 5 may be seen as $3+2$, $2+3$, $6-1$, and so on. Algebraic expressions are also procepts. The expressions $2(x+3)$ and $2x+6$ involve different processes of calculation, but they later give rise to the same function $f(x)=2(x+3)=2x+6$.

The steady progress through the curriculum in which different procedures are seen to be "essentially the same" is dealt with mathematically by introducing the notion of "equivalence." Early in the curriculum, fractions that involve different procedures giving the same final quantity such as $\frac{2}{4}$ and $\frac{3}{6}$ are said to be equivalent; later, they are conceived as being the same rational number.

Reflecting on this phenomenon that arises throughout arithmetic and algebra, I realized that even though we may distinguish between objects that are equivalent, mentally we operate more efficiently by thinking of them as being fundamentally the same object.

In Tall (2011), I extended the idea of flexible meaning of symbolism to the full range of mathematics. A "crystalline concept" was given a working definition as "a concept that has an internal structure of constrained relationships that cause it to have a necessary property as a consequence of its context."

In geometry, crystalline concepts include objects such as a circle which is defined to be the locus of a point that remains a fixed distance from its center, but has many constrained relationships (such as "the angle in a semicircle is a right angle"). In arithmetic, procepts are special cases of crystalline concepts, including various kinds of number such as whole numbers, fractions, signed numbers, real numbers, complex numbers, vectors, and expressions in algebra. In axiomatic mathematics, formal concepts defined set-theoretically are crystalline concepts whose properties are deduced from the definitions by mathematical proof.

Mathematicians use the notion of "equivalence relation" to deal with this idea in a technical way. For example, a fraction m/n may be considered as an ordered pair (m,n) of whole numbers under the equivalence relation

$$(m,n) \sim (p,q) \text{ if and only if } mq = np.$$

Then a rational number can be defined as an equivalence class of such pairs. At this stage, it is customary to agree that we can think of the equivalence class as a rational number. The notion of crystalline concept represents how we think about such mathematical ideas in practice, not as "equivalent" objects or as being "essentially" the same, but as *a single idea* that can be imagined in different ways.

The notion of a crystalline concept enables us to link together different kinds of representation in a single entity. For example, the real numbers may be defined as a complete ordered field, but it is possible to prove that two systems satisfying the axioms for a complete ordered field must be isomorphic and that they may be embodied (as a number line) and symbolized (as decimals). This gives the real numbers a rich structure as a crystalline concept which can be defined formally, visualized as points on a line, and operations can be performed using decimal arithmetic.

In other cases, a list of axioms, such as those for a group, may apply to many different examples, yet they all have a common structure given by the group axioms. We can think of a group as a crystalline concept. It is possible to show that the group structure may be embodied (as permutations of a set) or that groups may be classified in ways that may be proved from the axioms.

More generally, formal axiomatic systems often can be proved to satisfy "structure theorems" that classify its structure in a way that may often be embodied or symbolized in a manner that is more easily handled by the human mind. The notion of "crystalline concept" proves to be of value throughout the whole of mathematics, linking formal, embodied, and symbolic ways of thinking within an overall framework.

In school mathematics, these higher-level structures are not an explicit part of the curriculum. However, it is possible for learners to sense these structures at successive stages of development and for the teacher as mentor to encourage the learner to become aware of the underlying flexible structure. This includes an awareness of the ways in which concepts that are different at one stage may be classified at a later stage as one and the same.

In geometry, this development is formulated in terms of successive van Hiele levels. For instance, at the first visual level, squares and rectangles are seen as being different, but at the next level, a square is a special case of a rectangle. Likewise in arithmetic and algebra, expressions that represent different processes are later seen as representing the same crystalline concept. As the number systems become more sophisticated through whole numbers, fractions, signed numbers, infinite decimals, real numbers, and complex numbers, the crystalline structure subtly changes.

In practice, new ideas are often introduced by learning how to carry out procedures. For example, in the United States, the acronym "FOIL" is introduced to calculate $(a+b)(c+d)$ by multiplying the **f**irst elements in the brackets $a \times c$, then the **o**utside elements $a \times d$, then the **i**nside elements $b \times c$, then the **l**ast elements $b \times d$. We also teach a more subtle technique to factorize an expression such as x^2+5x+6 by seeking two numbers whose product is 6 and sum is 5.

This may have the unintended consequence that what is happening is the translation of one expression into a *different* expression, without realizing that these are just different ways of representing the same underlying crystalline concept. In this case, level reduction has occurred in which the student has learnt to carry out the procedures without grasping the rich flexibility of the mathematical structure.

In the United States, there are many examples of level reduction in teaching college algebra where textbooks are laid out using various devices such as color coding text or placing significant statements in boxes to remind the learner what should be remembered to be able to pass the test. As a result, there are more and more disconnected ideas to be remembered that increase the longer-term likelihood of overload and error.

A positive strategy is therefore to seek to make sense of the mathematics not only in terms of specific examples in practical contexts but also to draw out the pure thought that underlies the fundamental crystalline concepts.

This need not be a highly sophisticated activity. It simply means that the learner is encouraged to develop a sense of the underlying structures that cause mathematics to fit together in a coherent manner.

This requires more than a simple constructivist approach that encourages children to construct their own methods, for this may lead the learner investing effort into a particular way of working that becomes problematic at a later stage. Instead it requires the teacher, as mentor, to encourage the learner to seek to develop more powerful techniques that support longer-term learning while at the same time sensing the underlying crystalline structures that make mathematical thinking both more powerful and at the same time more flexible.

3.9 Supportive and Problematic Aspects of Individual Growth

Grasping the essential underlying ideas is not as simple as it sounds. While the teacher may be focusing on supportive aspects that generalize to new situations, the learner may be sensing problematic aspects that impede progress. The problem is that different students are affected in different ways by their own personal met-befores. This requires the teacher to be sensitive to problematic met-befores that may occur in some students but not in others so that different difficulties may be addressed in individual students or in groups of students that share a common conception.

For instance, in learning to solve equations, more successful students may realize that "doing the same thing to both sides" maintains the equality and so this proves to be a successful strategy. However, we have seen that many students are affected by problematic aspects that impede their progress. These may include ideas that have accumulated over many years and become more difficult to address as the ideas become deeply ingrained.

For a few students, algebra is essentially simple, even trivial. All one does is to use letters to represent numbers and manipulate them by the same methods with the same general properties noted in arithmetic. Others who are fixed in procedures without a flexible sense of relationships in arithmetic are more likely to find algebra becoming successively more complicated and even impossible, leading to a sense of anxiety that paralyzes thought and prevents future development.

3.10 The Long-Term Effects of Sense Making in Mathematical Thinking

Algebraic thinking is part of a long development from practical arithmetic with whole numbers, through more sophisticated number systems where general properties of arithmetic may be sensed and form a basis for the manipulation of symbols in algebra. As new ideas are encountered in fractions, signed numbers, generalized arithmetic, and algebra, the crystalline structure of the systems changes in ways that involve supportive aspects that generalize and problematic aspects that impede generalization.

Teaching that emphasizes the supportive ideas of generalization, as specified in many curricula around the world, will work with learners who grasp the flexible structure of successive contexts. These are the lucky ones who find mathematics pleasurable and powerful. But children who are impeded by problematic met-befores are less fortunate. If they cannot grasp the flexible structure, they may resort to learning procedurally to gain the pleasure of passing tests. In an era where testing at successive stages is prized, teachers and learners together may use a form of level reduction to know how to complete a specific task without setting it in a context that is appropriate for building meaningfully at the next stage. The consequence is that much teaching in our mathematics classrooms may be aimed at procedural competence rather than mathematical flexibility. In the longer term, successive level reduction impedes the development of flexible mathematical thinking.

However, not all individuals are the same or require the same level of mathematical sophistication. Is it sensible to make all students aspire to the same successive list of targets? If some students are finding mathematics difficult, repeating the same materials a second time may not necessarily help them think in the same way as those who regard mathematical ideas in more flexible ways. Different forms of employment in society require different forms of mathematical competence, and it is surely appropriate to teach children according to their needs and to help them learn mathematics in a way that makes sense to them.

Current curriculum development in many countries around the world is focusing on the introduction of problem solving and other aspects of mathematical communication, in lesson study in Japan, in realistic mathematics in the Netherlands, and in cooperative learning in the United States. All of these encourage positive achievement, while often remaining relatively silent on the negative.

Will the current drive to "raise standards" be successful? Time will tell. However, it is clear that teachers who have emotional difficulties with mathematics are likely to pass on their feelings to many of their students, so the desire to improve the positive aspects of learning for the students must be preceded by improvement of security and confidence of teachers, and this in turn requires the understanding of the broader aspects of mathematical thinking to be grasped by those who prepare the curriculum and set the tests.

The analysis of this chapter suggests that there are subtle underlying factors in the development of mathematical thinking that promote flexible thinking on the one hand and impede long-term learning on the other. This suggests a significant rethink in how we view the development of mathematical thinking. While an understanding of the increasingly sophisticated nature of crystalline concepts will help us to see the positive growth of supportive aspects of mathematical knowledge, an awareness of the problematic aspects carried forward from previous experience will help us assist individuals to deal with their difficulties that may become mathematical anxiety.

This suggests the need for mathematicians, curriculum designers, teachers, and learners to become explicitly aware of the underlying supportive and problematic aspects of long-term learning. It requires a global rethinking of the whole development of mathematical sense making that balances the subtle changes of the meaning of mathematics over the long term with the developing needs of different learners, building from early sense making in the perception and operation of the child and developing increasing sophistication of mathematical reasoning appropriate for differing roles in wider society. In particular, this suggests the development of a whole new course in teacher preparation that addresses not only the supportive aspects to deliver the positive objectives in long-term mathematical sense making while becoming aware of the specific problematic met-befores that impede student learning.

References

Burns, M. (1998). *Math: Facing an American phobia*. Sausalito, CA: Math Solutions Publications.

Clement, J., Lochhead, J., & Monk, G. S. (1981). Translation difficulties in learning mathematics. *American Mathematical Monthly, 88*, 286–290.

Gray, E. M., & Tall, D. O. (1994). Duality, ambiguity and flexibility: A proceptual view of simple arithmetic. *The Journal for Research in Mathematics Education, 26*(2), 115–141.

Hiebert, J., & Lefevre, P. (1986). Procedural and conceptual knowledge. In J. Hiebert (Ed.), *Conceptual and procedural knowledge: The case of mathematics* (pp. 1–27). Hillsdale, NJ: Erlbaum.

Lakoff, G., & Johnson, M. (1980). *Metaphors we live by*. Chicago: University of Chicago Press.

Lima, R. N. de. (2007). *Equações Algébricas no Ensino Médio: uma jornada por diferentes mundos da Matemática*. Ph.D. thesis, PUC/SP.

Lima, R. N., & Tall, D. O. (2008). Procedural embodiment and magic in linear equations. *Educational Studies in Mathematics, 67*(1), 3–18.

McGowen, M., & Tall, D. O. (2010). Metaphor or met-before? The effects of previous experience on the practice and theory of learning mathematics. *The Journal of Mathematical Behavior, 29*, 169–179.

McGowen, M., & Tall, D. O. (2013). Flexible thinking and met-befores: Impact on learning mathematics, with particular reference to the minus sign. *The Journal of Mathematical Behavior, 32*, 527–537.

NCTM. (2004). *Executive study principles and standards for school mathematics*. Retrieved May 24, 2015, from https://www.nctm.org/uploadedFiles/Standards_and_Positions/PSSM_ExecutiveSummary.pdf.

NCTM. (2013). *Common core state standards for mathematics*. Retrieved May 24, 2015, from http://www.nctm.org/ccssm/.
NGA. (2009). *Common core state standards development work group and feedback group announced*. Retrieved from http://www.nga.org/cms/home/news-room/news-releases/page_2009/col2-content/main-content-list/title_common-core-state-standards-development-work-group-and-feedback-group-announced.html.
Pisa (2012). Retrieved from http://www.oecd.org/pisa/keyfindings/pisa-2012-results.htm.
Schneider, M. (2014). *Who are the 24 people who wrote the common core standards*. Retrieved from http://dianeravitch.net/2014/04/28/mercedes-schneider-who-are-the-24-people-who-wrote-the-common-core-standards/.
Sfard, A. (2008). *Thinking as communicating*. New York: Cambridge University Press.
Skemp, R. R. (1976). Relational understanding and instrumental understanding. *Mathematics Teaching, 77*, 20–26.
Tall, D. O. (2004). Thinking through three worlds of mathematics, In *Proceedings of the 28th conference of the international group for the psychology of mathematics education, Bergen, Norway* (vol. 4, pp. 281–288).
Tall, D. O. (2011). Crystalline concepts in long-term mathematical invention and discovery. *For the Learning of Mathematics, 31*(1), 3–8.
Tall, D. O. (2013). *How humans learn to think mathematically*. New York: Cambridge University Press.
Tall, D. O., Lima, R. N., & Healy, L. (2014). Evolving a three-world framework for solving algebraic equations in the light of what a student has met before. *The Journal of Mathematical Behavior, 34*, 1–13.
Van Hiele, P. M. (1955). De niveau's in het denken, welke van belang zijn bij het onderwijs in de meetkunde in de eerste klasse van het VHMO Paed. In *Paedogogica studien XXXII* (pp. 289–297). Groningen: J. P. Wolters.
Van Hiele, P. M. (1986). *Structure and insight*. Orlando: Academic.

Chapter 4
Misconceptions and Learning Algebra

Julie L. Booth, Kelly M. McGinn, Christina Barbieri, and Laura K. Young

Abstract Rather than exclusively focus on mastery of procedural skills, mathematics educators are encouraged to cultivate conceptual understanding in their classrooms. However, mathematics learners hold many faulty conceptual ideas—or misconceptions—at various points in the learning process. In the present chapter, we first describe the common misconceptions that students hold when learning algebra. We then explain why these misconceptions are problematic and detail a potential solution with the capability to help students build correct conceptual knowledge while they are learning new procedural skills. Finally, we discuss other potential implications from the existence of algebraic misconceptions which require further study. In general, preventing and remediating algebraic misconceptions may be necessary for increasing student success in algebra and, subsequently, more advanced mathematics classes.

Keywords Misconceptions • Worked examples • Learning from errors • Conceptual knowledge • Self-explanation

4.1 Common Algebraic Misconceptions

Over the past several decades, researchers in mathematics education and educational psychology have identified a number of misconceptions that students tend to hold about algebraic content. While not an exhaustive list, a few of the most widely studied, including those dealing with equality/inequality, negativity, variables, fractions, order of operations, and functions, are discussed below.

J.L. Booth (✉)
Psychological Studies in Education, Temple University, 1301 Cecil B. Moore Avenue, Ritter Hall 239, Philadelphia, PA 19122, USA
e-mail: jlynne80@temple.edu

K.M. McGinn • C. Barbieri • L.K. Young
College of Education and Human Development, University of Delaware, Newark, DE USA

S. Stewart (ed.), *And the Rest is Just Algebra*, DOI 10.1007/978-3-319-45053-7_4

4.1.1 Equality/Inequality

Students at all levels have been found to hold misconceptions about the equal sign, including those enrolled in college calculus (Clement, Narode, & Rosnick, 1981). Often students have an operational understanding of the equal sign—the belief that the equal sign indicates where the answer should go—rather than a relational understanding, the belief that the equal sign indicates equivalence (Baroudi, 2006; Cheng-Yao, Yi-Yin, & Yu-Chun, 2014; Falkner, Levi, & Carpenter, 1999; Kieran, 1980, 1981; Van Dooren, Verschaffel, & Onghena, 2002). For example, of 375 sixth and seventh grade students, 58 % gave definitions for the equal sign that insinuated that the equal sign connects the answer to the problem (operational understanding), while only 29 % gave definitions that insinuated that the equal sign shows that what is to the left and the right of the sign mean the same thing (relational understanding) (Knuth, Alibali, Hattikudur, McNeil, & Stephens, 2008). While this type of arithmetic thinking may be sufficient during the early years, it causes major problems once students are asked to think algebraically (Booth & Koedinger, 2008; Knuth, Stephens, McNeil, & Alibali, 2006). Having a correct understanding of the meaning of the equal sign is imperative in order to manipulate and solve algebraic equations (Carpenter, Franke, & Levi, 2003; Kieran, 1981).

Some children believe that the equal sign cannot be used in an equation that does not have an operator symbol (i.e., $3=3$). These same students also believe that all operators must be on the left side of the equal sign. For instance, $5+2=3+4$ should be rewritten as $5+2=7$ and $3+4=7$ (Behr, Erlwanger, & Nichols, 1980). Furthermore, younger students tend to believe that the number immediately to the right of the equal sign must be the answer (Alibali, 1999; Falkner et al., 1999; Li, Ding, Capraro, & Capraro, 2008). For instance, in one particular study, all 145 sixth grade students incorrectly completed with number sentence $8+4=\_\_\_\_+5$ by filling in a 12 or 17 (Falkner et al., 1999). A second study found that about 76 % of 105 sixth graders were unable to correctly complete the first blank in the number sentence, $\_\_\_\_+3=5+7=\_\_\_\_$; however only about 13 % of those students were unable to answer the second (Li et al., 2008).

A similar misconception is one surrounding the concept of inequality. Similar to the equal sign, students at all levels tend to have difficulties with inequalities (Rowntree, 2009). Some students treat inequalities as equalities (Blanco & Garrote, 2007; Vaiyavutjamai & Clements, 2006). Others have a narrow understanding of the terms *more* or *less* (Warren, 2006). Finally, some students have major difficulties interpreting inequality solutions (Tsamir & Bazzini, 2004; Vaiyavutjamai & Clements, 2006).

4.1.2 Negativity

Another category of algebraic misconceptions is dealing with negativity. Those with an incorrect or incomplete understanding of the negative sign are more likely to use incorrect strategies when solving algebraic equations (Booth & Koedinger, 2008). Due to the abstract nature of negativity, this concept is especially difficult for students moving from arithmetic to algebraic thinking (Linchevski & Williams, 1999). These students tend to only link the negative sign with the binary operation of subtraction. For instance, Vlassis (2002, 2004) found that most eighth graders can easily interpret the meaning of negative nine within the expression $n - 9$, but have trouble when −9 is presented alone.

Difficulties with the negative sign persist into the college years. Cangelosi, Madrid, Cooper, Olson, and Hartter (2013) found that college students have difficulty manipulating exponential expressions when a negative sign is included as part of the base, preceding the base, or as part of the exponent. For instance, students often misinterpret $-9^{3/2}$ as $(-9)^{3/2}$ (Cangelosi et al., 2013).

4.1.3 Variables

Misconceptions dealing with the use of variables are also widely studied. One of the more common misunderstandings is the belief that the letter in a number sentence stands for an actual object or is a label (Asquith, Stephens, Knuth, & Alibali, 2007; Clement, 1982; MacGregor & Stacey, 1997; McNeil et al., 2010; Stacey & MacGregor, 1997; Usiskin, 1988). This misinterpretation can be seen in the classic error to the "student and professor" problem. When students are asked to write a number sentence to represent the phrase, six times as many students as professors, the most common error is 6s = p (Clement, Lochhead, & Monk, 1981; Rosnick, 1981). Students believe that s was a label for students, rather than a variable representing the number of students (Rosnick, 1981).

Alternatively, some students will ignore the variables altogether. For instance, when asked to solve $(n + 5) + 4$, 20 % of students incorrectly give the answer of 9, ignoring the n (Kuchemann, 1978). Others believe that the letter is associated with its position in the alphabet (Asquith et al., 2007; Herscovics & Kieran, 1980; MacGregor & Stacey, 1997; Watson, 1990). Furthermore, students have trouble understanding that the same letter seen multiple times in a number sentence must represent that same number (Kieran, 1985) or that different letters within a number sentence can also represent the same number (Stephens, 2005; Swan, 2000). On a similar note, students also often misunderstand the meaning of operational symbols when paired with variables. For instance, since students are used to joining two terms when they see the addition symbol (i.e., 2 + ½ = 2½), they will mistakenly believe that $2 + x$ is the same as $2x$ (Booth, 1986).

4.1.4 *Fractions*

While introduced before algebraic concepts, fraction misconceptions also greatly influence students' acquisition of algebra knowledge. The National Mathematics Advisory Panel (2008) suggests that one of the most important types of knowledge necessary for algebra learning is knowledge of rational numbers or fractions. Fractions can be seen in algebra as coefficients/slope, constants, and solutions (Wu, 2001). Brown and Quinn (2006) assessed Algebra I students' fraction knowledge and found that students have trouble writing a fraction to represent the shaped part of a figure, simplifying fractions to lowest terms, adding and subtraction fractions, and multiplying and dividing fractions. Specifically, students often misused the cross multiplying algorithm when attending to multiply fractions, failed to use the inverse operations to solve equations, and failed to even attempt the problem.

4.1.5 *Order of Operations*

Another type of misconception affecting students of all ages deals with the order of operations and use of brackets (Kieran, 1985; Pinchback, 1991). Many students do not see the need to adhere to the order of operations rules and resort to solving the expression from left to right (Gardella, 2009; Kieran, 1979). Furthermore, many students fail to realize that brackets can be used to both groups together as well as signal multiplication (i.e., $(20-7)=13$ and $-(20-7)=-13$; Linchevski, 1995).

4.1.6 *Functions*

Lastly, students often misinterpret the meaning of algebraic functions. For instance, some students treat a graph as a picture of a given scenario (i.e., a graph comparing speed and time) (Clement, 1989). Furthermore, both students and adults tend to believe that a linear function must be proportional simply because it increases or decreases at a constant rate (Pugalee, 2010; Van de Walle, Karp, & Bay-Williams, 2013); however this is only true when the function passes through the origin.

4.2 Why Should We Be Concerned About Misconceptions?

The previous section described a number of misconceptions students tend to have when learning algebra. It is well established and documented that such misconceptions exist. But why is having these misconceptions problematic? In this section, we describe a number of ways in which having misconceptions, or flawed conceptual knowledge of algebra, might impact students' performance and learning.

4.2.1 Relation to Procedural Skills

Having good procedural skills, or the ability to carry out procedures to solve problems (Rittle-Johnson, Siegler, & Alibali, 2001), is arguably a critical component of success in mathematics (Kilpatrick, Swafford, & Findell, 2001). It has been well established that conceptual knowledge and procedural skill are related (Rittle-Johnson & Siegler, 1998), and some researchers maintain that the two in fact fall on a single continuum (Star, 2005). Though the two develop iteratively and one or the other may come first depending on the particular content (Rittle-Johnson & Siegler, 1998), for many mathematics domains, it is necessary to have correct conceptual knowledge in order to develop correct procedural skills.

Work in algebra has established that students with stronger conceptual knowledge are better at solving equations and are able to learn new procedures more easily than peers with flawed conceptual knowledge (e.g., Booth, Koedinger, & Siegler, 2007; Sweller & Cooper, 1985). In particular, students who hold misconceptions about the equal sign or negative signs solve fewer equations correctly and have greater difficulty learning how to solve equations (Booth & Koedinger, 2008). Correction of these misconceptions can lead to improvements in equation-solving skills (Booth & Koedinger, 2008).

4.2.2 Relation to Problem Encoding

The ability to correctly encode a problem, or perceptually process the important features of the problem and create an internal representation that can be used later (Chase & Simon, 1973), has been repeatedly shown to be important for problem-solving success (Alibali, Phillips, & Fischer, 2009; Booth & Davenport, 2013; Rittle-Johnson & Alibali, 1999; Siegler, 1976). Prior knowledge necessarily impacts how a learner encodes a problem. For example, students are better at encoding equations that are familiar and tend to misencode problem features in unfamiliar equations as if they follow the structure of more familiar problems (McNeil & Alibali, 2004).

Conceptual knowledge also impacts learners' encoding of problems. Experts in a domain encode problems more accurately than novices (Chase & Simon, 1973; Chi, Feltovich, & Glaser, 1981), and algebra students with more correct conceptual knowledge have been shown to have higher encoding accuracy (Booth & Davenport, 2013). This is, perhaps, not surprising, as correct encoding requires noticing the important features in a problem and conceptual knowledge helps students determine what features are important (Crooks & Alibali, 2013;

Prather, 2012; Rittle-Johnson & Alibali, 1999). In other words, when students have flawed conceptual knowledge, they may not be able to correctly determine which features to focus on and/or may not consider those features in a meaningful way (Booth & Davenport, 2013).

4.2.3 *Relation to Specific Problem-Solving Errors*

4.2.3.1 Misconceptions and Related Errors

Algebraic misconceptions that students hold predict the types of errors students make during problem-solving (Booth & Koedinger, 2008). Durkin and Rittle-Johnson (2015) demonstrate that errors made with high confidence during problem-solving are representative of strongly held misconceptions that are more difficult to overcome with instruction. Oftentimes, these errors arise when students are learning a new topic and attempt unsuccessfully to relate it to something they've learned prior. Although this can sometimes be a useful strategy, when rules or strategies are overgeneralized, this can lead to struggles as well (Stagylidou & Vosniadou, 2004; Vamvakoussim & Vosniadou, 2004), making students particularly resistant to conceptual change in mathematics (McNeil, 2014).

One common example of when students struggle to learn and apply altering rules during problem-solving when moving to higher levels of mathematics is when they transition from dealing with solely natural numbers to all rational numbers (Van Dooren, Lehtinen, & Vershcaffel, 2015). A natural number bias can often lead students to make errors when dealing with fractions and decimals. Another is when students are asked to understand and use the equal sign as a symbol or equivalence between two expressions in algebra rather than the more commonly used form of seemingly signaling that the student should carry out an operation. This can often lead students to making the error of performing the given operation on all given numbers, regardless of where the numbers are located within the equation (McNeil & Alibali, 2004).

Errors that persist are often an indication that a student holds an underdeveloped understanding of a particular underlying concept (Cangelosi et al., 2013). Analyzing errors that students make during problem-solving is one useful method for learning more about the particular misconceptions that students hold (Clement, 1982; Corder, 1982; Liebenburg, 1997).

4.2.3.2 Persistence of Errors

Certain errors that students make in mathematics are quite persistent and lead to troubles at different levels of mathematics. Most of the misconceptions addressed within this chapter are expressed in algebra. It is vital to understand these misconceptions as Algebra I is considered a gatekeeper course to higher-level STEM courses (Adelman, 2006). However, understanding of algebra is arguably built upon early arithmetic knowledge (Bodin & Capponi, 1996), so it is important to consider how misconceptions in earlier stages of mathematics can lead to errors made later on in algebra. For example, Mazzocco and colleagues (Mazzocco, Murphy, Brown, Rinne, & Herold, 2013) found that errors made in second and third grade are predictive of not only specific types of errors made in eighth grade but also speed during problem-solving. Specifically, students who made particular errors in a symbolic number task in second or third grade were slower and made more errors when completing addition and multiplication computations in eighth grade.

Algebra I is most commonly taken in the eighth or ninth grade. However, some errors made during secondary mathematics have been found to persist even into postsecondary levels of mathematics. Negative sign errors have been found to be quite common and quite persistent at varying levels of mathematics (Booth, Barbieri, Eyer, & Paré-Blagoev, 2014; Seng, 2010). Being able to manipulate integers is a subordinate skill in algebra and higher levels of mathematics. Therefore, it is clear as to why misconceptions about the negative sign (as well as the equal sign) have been found to interfere with students' learning of how to solve algebraic equations (Booth & Koedinger, 2008). This applies to students who may stereotypically be considered advanced or students who manage to complete school standards for Algebra I as well. Negative sign errors are common and interfere with learning at varying levels of mathematics (Kieran, 2007). In a cross-sectional study, Cangelosi and colleagues found that negative sign errors made in College Algebra (e.g., incorrectly simplifying negative numbers with a rational exponent) persist through Calculus II (Cangelosi et al., 2013).

4.2.3.3 Relation of Errors to Learning

Conceptual change is undoubtedly a slow and gradual process (McNeil & Alibali, 2005; Vamvakoussi & Vosniadou, 2010). While some misconceptions seem to persist as demonstrated in the errors students make all the way through college, other misconceptions change in prevalence and persistence based upon the content to be learned (Booth et al., 2014). For example, Durkin and Rittle-Johnson (2015) explored changes in misconceptions when judging the magnitude of decimals over the course of a 1-month period of instruction. *Whole number errors* and *role of zero errors* started off prevalent but declined over time. Whole number errors were classified as those that indicate treating a decimal as if they are whole numbers and believing more numbers to the right of the decimal means a larger number. The role of zero errors were classified as those that indicate treating a decimal with a zero in the tenths place as if the following digit is actually in the tenths place. These errors were considered to be representations of a whole number bias (Ni & Zhou, 2005). However, fraction errors, in which students try to relate the length of the decimal to its magnitude, increased over time. Durkin and Rittle-Johnson suggest that this change in prevalence of types of errors indicates change in conceptual thinking about number. However, how the prevalence and persistence of these errors predicted later achievement was not addressed.

Booth and colleagues (Booth et al., 2014) conducted a similar analysis upon errors over the course of an academic year and found that making certain types of errors while learning particular content in algebra is indicative of detriment to mathematics achievement. For example, students who made variable errors at the beginning of the academic year while taking Algebra I on arguably what would be simpler content demonstrated lower mathematics achievement scores at the end of the academic year. Students who made more errors related to mathematical properties (i.e., inappropriately applying the distributive, commutative, or associative properties) or who conducted the wrong operations during the beginning and middle of the year also struggled on the end of year achievement test. Students who made more errors involving equality and inequality at the middle and end of year also demonstrated lower achievement. Lastly, negative sign and arithmetic errors at the end of the year, when content was presumably most difficult, were indicative of low mathematics achievement. Results from this study emphasize the importance of considering how errors stemming from misconceptions align with particular content. Understanding not only the prevalence and persistence of mathematical errors in relation to particular content but also what these errors indicate about the misconceptions students hold and how these impact future learning are vital first steps when considering designing appropriate interventions that address student misconceptions.

4.3 How Can We Address Student Misconceptions?

A number of interventions exist which aim to improve students' conceptual understanding in algebra, including those focused on reteaching fundamental concepts and principles (Ma, 1999), having students compare multiple solution methods (Rittle-Johnson & Star, 2007), or completely reforming mathematics curricula to be contextualized in real-world problems (Hiebert et al., 1996) or conceptual models (Xin, Wiles, & Lin, 2008). In this chapter, we describe one particular method which has proved to be effective at both improving student's conceptual understanding and procedural skill in algebra. This approach stems from three scientific principles on how people learn: self-explanation, worked examples, and cognitive dissonance. Each of these three principles is described below, before we explain how they have been combined and review findings on the effectiveness of this combination.

4.3.1 Self-Explanation

Self-explanation is defined as explaining information to oneself while reading or studying (Chi, 2000). Early evidence revealed that better learners do this naturally (Chi, Bassok, Lewis, Reimann, & Glaser, 1989), and follow-up studies examined the effectiveness of prompting all students to explain. The self-explanation principle maintains that there are a number of benefits for learning when students are asked to explain information to themselves while reading or studying (Chi, 2000). Some of these benefits include improvement in the degree to which students integrate new information with their prior knowledge, make the newly learned knowledge explicit, and, subsequently, notice gaps in their knowledge and draw inferences to fill those gaps (Chi, 2000; Roy & Chi, 2005).

4.3.2 Worked Examples

Traditional instruction, particularly in science, technology, engineering, and mathematics (STEM) domains, involves demonstrating the procedures for solving problems (on the blackboard, on the smart board, in the textbook) and then having students practice solving those types of problems on their own. However, a large body of work from laboratory studies suggests that these worked examples should not just occur at the beginning of the lesson—they should be interleaved within the practice sessions as well (e.g., Cooper & Sweller, 1987; Sweller & Cooper, 1985; Trafton & Reiser, 1993). The worked example principle maintains that replacing some (or even half) of the practice problems with worked-out solutions for students to study can increase learning of the procedures to solve problems, even though the students have less practice solving those problems themselves (Sweller, 1999). Benefits of focusing students' limited cognitive capacities on understanding the concepts and procedures necessary for problem-solving (rather than on attempting to apply procedures by rote) include faster mastery of instructed procedures (Clark & Mayer, 2003; Schwonke et al., 2009) and increased transfer of procedural skills to solve more difficult problems (Catrambone, 1996, 1998; Cooper & Sweller, 1987).

4.3.3 Cognitive Dissonance

The idea of cognitive dissonance stems from a theory purported by Festinger (1957), which maintains that humans naturally seek consistency between their beliefs and the reality observed in the world and that a clash between belief and reality leads to an unpleasant feeling and a drive to resolve the discrepancy. In other words, if one is presented with information that conflicts with their own beliefs, they will work to make sense of the differences so they can return to a harmonious state. Creating such cognitive disequilibrium is thus proposed to be an effective technique for producing change in thinking (e.g., Graesser, 2009).

One method of promoting cognitive dissonance is through the presentation of errors for students to consider and study. Learning from errors is thought to be effective because it prompts students to identify features of problems that make the demonstrated procedure incorrect, which in turn can help students correct their own misconceptions (Ohlsson, 1996). An additional benefit of studying and explaining errors is that it may help learners acknowledge that the demonstrated procedure is wrong and make it less likely they will utilize that procedure themselves when solving problems (Siegler, 2002).

4.3.4 Combining Self-Explanation, Worked Examples, and Learning from Errors

These principles, which have been well tested in laboratory settings, have been combined into a single effective intervention: explaining correct and incorrect worked examples during problem-solving practice. Essentially, for some of the items in practice assignments, students are shown an example of a fictitious learner's problem solution—solved either correctly or incorrectly and clearly marked as such—and asked to explain the example in response to one or more prompts about particular features in the problems, about particular errors made in solutions, or about how the fictitious learner might be thinking about the problem.

Prior research had established that, compared to studying correct worked examples, *explaining* correct examples increased students' conceptual knowledge (Hilbert, Renkl, Schworm, Kessler, & Reiss, 2008) and their ability to solve both similar and more difficult problems (Renkl, Stark, Gruber, & Mandl, 1998). Further research suggested that explaining correct *and incorrect examples* further increased learning benefits for building correct conceptual understanding (Adams et al., 2014; Booth et al., 2015; Booth, Lange, Koedinger, & Newton, 2013) and decreasing student misconceptions (Durkin & Rittle-Johnson, 2012). Recently, in a randomized controlled trial in real-world classrooms across an entire Algebra 1 curriculum, this combination led to robust improvements on conceptual and procedural skills as well as skills specifically measured by standardized achievement tests (Booth et al., 2015); benefits for conceptual understanding were even stronger for students who were struggling with the material (Booth, Oyer, et al., 2015).

4.4 Practical Implications of the Existence and Persistence of Algebraic Misconceptions

By now, we can hopefully agree that algebraic misconceptions are a problem and that traditional algebra instruction is not doing enough to remedy the problem. We have offered one suggestion of how to change algebra instruction to better target and fix student misconceptions and allow them to move forward productively with learning more difficult algebraic content. This is certainly not the only option for how to alter algebra instruction; any interventions geared toward improving conceptual understanding (while still building procedural skill) may be good candidates for instruction.

However, full remediation may require looking backward as well. Misconceptions don't typically develop out of the blue; they develop them because children are trying to make sense of the world around them by using the information made available (Vosniadou & Brewer, 1992). What information are we making available in younger grades that lead to students developing algebraic misconceptions? One line of work suggests that the way we teach earlier math can have a profound effect on students' understanding of algebraic concepts. For example, McNeil and Alibali (2005) showed that elementary school students' knowledge of arithmetic operation patterns (e.g., operations = answer) hinders their ability to learn from a lesson on solving equations; unfortunately, mathematics textbooks rarely present the equal sign in a context that would encourage a relational understanding—most presentations are the standard operations = answer format (e.g., $6+2=8$) that hinders learning (McNeil et al., 2006). Giving children more practice, solving problems in this format also makes it less likely that they will build a correct concept of mathematical equivalence (McNeil, 2008). One could imagine similar consequences for early presentations and practice (or lack thereof) with negative signs and variables.

How can we prevent such ingrained misconceptions from developing? One possibility may be a combination of systemic changes to early mathematics instruction and materials and the approach described in this chapter. We must change the way we introduce algebraic problem features and concepts in the first place. Recent recommendations stress focusing on such concepts earlier in the mathematics curriculum (e.g., CCSSI, 2010). We must always think about how we are presenting information to young children and whether it will help them build a correct concept. Second, teachers can have students explain correct and incorrect worked examples in earlier grades to help them focus on building a correct conceptual foundation as well as the necessary procedural skills. This may help prevent formation and entrenchment of these misconceptions early on. By preventing and/or quickly remediating misconceptions, we can help future generations have a smoother transition to—and greater success in—learning algebra.

Acknowledgements Funding for the writing of this chapter was provided by the Institute of Education Sciences and U.S. Department of Education through Grant R305B150014 to Temple University, Grant R305B130012 to the University of Delaware, and Grant R305A100150 to the Strategic Education Research Partnership. The opinions expressed are those of the authors and do not represent views of the Institute or the U.S. Department of Education.

References

Adams, D. M., McLaren, B. M., Durkin, K., Mayer, R. E., Rittle-Johnson, B., Isotani, S., et al. (2014). Using erroneous examples to improve mathematics learning with a web-based tutoring system. *Computers in Human Behavior, 36*, 401–411.

Adelman, C. (2006). *The toolbox revisited: Paths to degree completion from high school through college*. Washington, DC: US Department of Education.

Alibali, M. W. (1999). How children change their minds: Strategy change can be gradual or abrupt. *Developmental Psychology, 35*(1), 127–145.

Alibali, M. W., Phillips, K. M. O., & Fischer, A. D. (2009). Learning new problem-solving strategies leads to changes in problem representation. *Cognitive Development, 24*, 89–101.

Asquith, P., Stephens, A. C., Knuth, E. J., & Alibali, M. W. (2007). Middle school mathematics teachers' knowledge of students' under- standing of core algebraic concepts: Equal sign and variable. *Mathematical Thinking and Learning: An International Journal, 9*(3), 249–272.

Baroudi, Z. (2006). Easing students' transition to algebra. *Australian Mathematics Teacher, 62*(2), 28–33.

Behr, M. J., Erlwanger, S., & Nichols, E. (1980). How children view the equals sign. *Mathematics Teaching, 92*, 13–15.

Blanco, L. J., & Garrote, M. (2007). Difficulties in learning inequalities in students of the first year of pre-university education in Spain. *Eurasia Journal of Mathematics, Science and Technology Education, 3*, 221–229.

Bodin, A., & Capponi, B. (1996). Junior secondary school practices. In A. J. Bishop et al. (Eds.), *International handbook of mathematics education* (pp. 565–614). London: Kluwer Academic.

Booth, L. R. (1986). Difficulties in algebra. *Australian Mathematics Teacher, 42*(3), 2–4.

Booth, J. L., Barbieri, C., Eyer, F., & Paré-Blagoev, J. (2014). Persistent and pernicious errors in algebraic problem-solving. *The Journal of Problem Solving, 7*, 10–23.

Booth, J. L., Cooper, L. A., Donovan, M. S., Huyghe, A., Koedinger, K. R., & Paré-Blagoev, E. J. (2015). Design-based research within the constraints of practice: AlgebraByExample. *Journal of Education for Students Placed at Risk, 20*(1–2), 79–100.

Booth, J. L., & Davenport, J. L. (2013). The role of problem representation and feature knowledge in algebraic equation solving. *The Journal of Mathematical Behavior, 32*(3), 415–423.

Booth, J. L., & Koedinger, K. R. (2008). Key misconceptions in algebraic problem solving. In B. C. Love, K. McRae, & V. M. Sloutsky (Eds.), *Proceedings of the 30th annual cognitive science society* (pp. 571–576). Austin, TX: Cognitive Science Society.

Booth, J. L., Koedinger, K. R., & Siegler, R. S. (2007). [Abstract]. The effect of prior conceptual knowledge on procedural performance and learning in algebra. In D. S. McNamara & J. G. Trafton (Eds.), *Proceedings of the 29th annual cognitive science society*. Austin, TX: Cognitive Science Society.

Booth, J. L., Lange, K. E., Koedinger, K. R., & Newton, K. J. (2013). Using example problems to improve student learning in algebra: Differentiating between correct and incorrect examples. *Learning and Instruction, 25*, 24–34.

Booth, J. L., Oyer, M. H., Paré-Blagoev, E. J., Elliot, A., Barbieri, C., Augustine, A. A., et al. (2015). Learning algebra by example in real-world classrooms. *Journal of Research on Educational Effectiveness, 8*(4), 530–551.

Brown, G., & Quinn, R. J. (2006). Algebra students' difficulty with fractions. *Australian Mathematics Teacher, 62*(4), 28–40.

Cangelosi, R., Madrid, S., Cooper, S., Olson, J., & Hartter, B. (2013). The negative sign and exponential expressions: Unveiling students' persistent errors and misconceptions. *The Journal of Mathematical Behavior, 32*(1), 69–82.

Carpenter, T. P., Franke, M. L., & Levi, L. (2003). *Thinking mathematically: Integrating arithmetic and algebra in elementary school.* Portsmouth: Heinemann.

Catrambone, R. (1996). Generalizing solution procedures learned from examples. *Journal of Experimental Psychology. Learning, Memory, and Cognition, 22*(4), 1020–1031.

Catrambone, R. (1998). The subgoal learning model: Creating better examples so that students can solve novel problems. *Journal of Experimental Psychology, 127*(4), 355–376.

Chase, W. G., & Simon, H. A. (1973). Perception in chess. *Cognitive Psychology, 4*, 55–81.

Cheng-Yao, L., Yi-Yin, K., & Yu-Chun, K. (2014). Changes in pre-service teachers' algebraic misconceptions by using computer-assisted instruction. *International Journal for Technology in Mathematics Education, 21*(3), 21–30.

Chi, M. T. (2000). Self-explaining expository texts: The dual processes of generating inferences and repairing mental models. *Advances in Instructional Psychology, 5*, 161–238.

Chi, M. T., Bassok, M., Lewis, M. W., Reimann, P., & Glaser, R. (1989). Self-explanations: How students study and use examples in learning to solve problems. *Cognitive Science, 13*(2), 145–182.

Chi, M. T., Feltovich, P. J., & Glaser, R. (1981). Categorization and representation of physics problems by experts and novices. *Cognitive Science, 5*(2), 121–152.

Clark, R. C., & Mayer, R. E. (2003). *e-Learning and the science of instruction: Proven guidelines for consumers and designers of multimedia learning*. San Francisco, CA: Jossey-Bass.

Clement, J. (1982). Algebra word problem solutions: Thought processes underlying a common misconception. *Journal for Research in Mathematics Education, 13*(1), 16–30.

Clement, J. (1989). The concept of variation and misconceptions in Cartesian graphing. *Focus on Learning Problems in Mathematics, 11*, 77–87.

Clement, J., Lochhead, J., & Monk, G. (1981). Translation difficulties in learning mathematics. *American Mathematical Monthly, 88*, 286–290.

Clement, J., Narode, R., & Rosnick, P. (1981). Intuitive misconceptions in algebra as a source of math anxiety. *Focus on Learning Problems in Mathematics, 3*(4), 36–45.

Common Core State Standards Initiative. (2010). *Common core state standards for mathematics.* Retrieved from http://www.corestandards.org/assets/CCSSI_Math%20Standards.pdf.

Cooper, G., & Sweller, J. (1987). Effects of schema acquisition and rule automation on mathematical problem-solving transfer. *Journal of Educational Psychology, 79*(4), 347–362.

Corder, S. P. (1982). *Error analysis & interlanguage*. Oxford: Oxford University Press.

Crooks, N. M., & Alibali, M. W. (2013). Noticing relevant problem features: Activating prior knowledge affects problem solving by guiding encoding. *Frontiers in Psychology, 4*, 884. doi:10.3389/fpsyg.2013.00884.

Durkin, K., & Rittle-Johnson, B. (2012). The effectiveness of using incorrect examples to support learning about decimal magnitude. *Learning and Instruction, 22*(3), 206–214.

Durkin, K., & Rittle-Johnson, B. (2015). Diagnosing misconceptions: Revealing changing decimal fraction knowledge. *Learning and Instruction, 37*, 21–29.

Falkner, K. P., Levi, L., & Carpenter, T. P. (1999). Children's understanding of equality: A foundation for algebra. *Teaching Children Mathematics, 6*(4), 232–236.

Festinger, L. (1957). *A theory of cognitive dissonance*. Evanston, IL: Row Peterson.

Gardella, F. J. (2009). *Introducing difficult mathematics topics in the elementary classroom: A teacher's guide to initial lessons.* New York: Routledge, Taylor & Francis.

Graesser, A. C. (2009). Inaugural editorial for journal of educational psychology. *Journal of Educational Psychology, 101*(2), 259–261.

Herscovics, N., & Kieran, C. (1980). Constructing meaning for the concept of equation. *Mathematics Teacher, 73*, 572–580.

Hiebert, J., Carpenter, T. P., Fennema, E., Fuson, K., Human, P., Murray, H., et al. (1996). Problem solving as a basis for reform in curriculum and instruction: The case of mathematics. *Educational Researcher, 25*(4), 12–21.

Hilbert, T., Renkl, A., Schworm, S., Kessler, S., & Reiss, K. (2008). Learning to teach with worked-out examples: A computer-based learning environment for teachers. *Journal of Computer Assisted Learning, 24*(4), 316–332.

Kieran, C. (1979). Children's operational thinking within the context of bracketing and the order of operations. In *Paper presented at the third international conference for the psychology of mathematics education, Coventry, England.*

Kieran, C. (1980). Constructing meaning for non-trivial equations. In *Paper presented at the annual meeting of the American educational research association, Boston, MA.*

Kieran, C. (1981). Concepts associated with the equality symbol. *Educational Studies in Mathematics, 12*, 318–326.

Kieran, C. (1985). The equation-solving errors of novice and intermediate algebra students. In *Paper presented at the ninth international conference for the psychology of mathematics education, The Netherlands.*

Kieran, C. (2007). Learning and teaching of algebra at the middle school through college levels: Building meaning for symbols and their manipulation. In F. K. Lister (Ed.), *Second handbook of research on mathematics teaching and learning* (pp. 707–762). Reston, VA: NCTM.

Kilpatrick, J., Swafford, J., & Findell, B. (2001). Adding it up. In *Mathematics Learning Study Committee, Center for Education*. Washington, DC: National Academy Press.

Knuth, E. J., Alibali, M. W., Hattikudur, S., McNeil, N. M., & Stephens, A. C. (2008). The importance of equal sign understanding in the middle grades. *Mathematics Teaching in the Middle School, 13*(9), 514–519.

Knuth, E. J., Stephens, A. C., McNeil, N. M., & Alibali, M. W. (2006). Does understanding the equal sign matter? Evidence from solving equations. *Journal for Research in Mathematics Education, 37*, 297–312.

Kuchemann, D. (1978). Children's understanding of numerical variables. *Mathematics in School, 7*, 23–26.

Li, X., Ding, M., Capraro, M. M., & Capraro, R. M. (2008). Sources of differences in children's understandings of mathematical equality: Comparative analysis of teacher guides and student texts in China and the United States. *Cognition and Instruction, 26*(2), 195–217.

Liebenburg, R. (1997). The usefulness of an intensive diagnostic test. In P. Kelsall & M. de Villiers (Eds.), *Proceedings of the third national congress of the association for mathematics education of South Africa* (Vol. 2, pp. 72–79). Durban, South Africa: University of Natal.

Linchevski, L. (1995). Algebra with numbers and arithmetic with letters: A definition of pre-algebra. *The Journal of Mathematical Behavior, 14*, 113–120.

Linchevski, L., & Williams, J. (1999). Using intuition from everyday life in 'filling' the gap in children's extension of their number concept to include the negative numbers. *Educational Studies in Mathematics, 39*, 131–147.

Ma, L. (1999). *Knowing and teaching elementary mathematics: Teachers' understanding of fundamental mathematics in China and the United States*. Mahwah, NJ: Lawrence Erlbaum Associates.

MacGregor, M., & Stacey, K. (1997). Students' understanding of algebraic notation: 11–15. *Educational Studies in Mathematics, 33*(1), 1–19.

Mazzocco, M. M. M., Murphy, M. M., Brown, E. C., Rinne, L., & Herold, C. H. (2013). Persistent consequences of atypical early number concepts. *Frontiers in Psychology, 4*, 1–9.

McNeil, N. M. (2008). Limitations to teaching children $2+2=4$: Typical arithmetic problems can hinder learning of mathematical equivalence. *Child Development, 79*, 1524–1537.

McNeil, N. M. (2014). A change-resistance account of children's difficulties understanding mathematical equivalence. *Child Development Perspectives, 8*(1), 42–47.

McNeil, N. M., & Alibali, M. W. (2004). You'll see what you mean: Students encode equations based on their knowledge of arithmetic. *Cognitive Science, 28*(3), 451–466.

McNeil, N. M., & Alibali, M. W. (2005). Why won't you change your mind? Knowledge of operational patterns hinders learning and performance on equations. *Child Development, 76*(4), 883–899.

McNeil, N. M., Grandau, L., Knuth, E. J., Alibali, M. W., Stephens, A. C., Hattikudur, S., et al. (2006). Middle-school students' understanding of the equal sign: The books they read can't help. *Cognition and Instruction, 24*(3), 367–385.

McNeil, N. M., Weinberg, A., Hattikudur, S., Stephens, A. C., Asquith, P., Knuth, E. J., et al. (2010). A is for apple: Mnemonic symbols hinder the interpretation of algebraic expressions. *Journal of Educational Psychology, 102*(3), 625–634.

National Mathematics Advisory Panel. (2008). *Foundations for success: The final report of the national mathematics advisory panel*. Washington, DC: US Department of Education.

Ni, Y., & Zhou, Y.-D. (2005). Teaching and learning fraction and rational numbers: The origins and implications of whole number bias. *Educational Psychologist, 40*(1), 27–52.

Ohlsson, S. (1996). Learning from performance errors. *Psychological Review, 103*(2), 241.

Pinchback, C. L. (1991). Types of errors exhibited in a remedial mathematics course. *Focus on Learning Problems in Mathematics, 13*(2), 53–62.

Prather, R. W. (2012). Implicit learning of arithmetic regularities is facilitated by proximal contrast. *PLoS One, 7*(10), e48868. doi:10.1371/journal.pone.0048868.

Pugalee, D. (2010). Extending students' development of proportional reasoning. In *Paper presented at the regional meeting of the national council of teachers of mathematics, New Orleans, LA*.

Renkl, A., Stark, R., Gruber, H., & Mandl, H. (1998). Learning from worked-out examples: The effects of example variability and elicited self-explanations. *Contemporary Educational Psychology, 23*(1), 90–108.

Rittle-Johnson, B., & Alibali, M. W. (1999). Conceptual and procedural knowledge of mathematics: Does one lead to the other? *Journal of Educational Psychology, 91*(1), 175–189.

Rittle-Johnson, B., & Siegler, R. S. (1998). The relationship between conceptual and procedural knowledge in learning mathematics: A review. In C. Donlan (Ed.), *The development of mathematical skills* (pp. 75–110). East Sussex, England: Psychology Press.

Rittle-Johnson, B., Siegler, R. S., & Alibali, M. W. (2001). Developing conceptual understanding and procedural skill in mathematics: An iterative process. *Journal of Educational Psychology, 93*, 346–362.

Rittle-Johnson, B., & Star, J. R. (2007). Does comparing solution methods facilitate conceptual and procedural knowledge? An experimental study on learning to solve equations. *Journal of Educational Psychology, 99*(3), 561.

Rosnick, P. (1981). Some misconceptions concerning the concept of variable. *Mathematics Teacher, 74*, 418–420.

Rowntree, R. V. (2009). Students' understandings and misconceptions of algebraic inequalities. *School Science and Mathematics, 109*(6), 311–312.

Roy, M., & Chi, M. T. H. (2005). The self-explanation effect in multimedia learning. In R. E. Mayer (Ed.), *The Cambridge handbook of multimedia learning* (pp. 271–286). New York: Cambridge University Press.

Schwonke, R., Renkl, A., Krieg, C., Wittwer, J., Aleven, V., & Salden, R. (2009). The worked-example effect: Not an artefact of lousy control conditions. *Computers in Human Behavior, 25* (2), 258–266.

Seng, L. K. (2010). An error analysis of Form 2 (Grade 7) students in simplifying algebraic expressions: A descriptive study. *Electronic Journal of Research in Educational Psychology, 8* (1), 139–162.

Siegler, R. S. (1976). Three aspects of cognitive development. *Cognitive Psychology, 8*, 481–520.

Siegler, R. S. (2002). Microgenetic studies of self-explanations. In N. Granott & J. Parziale (Eds.), *Microdevelopment: Transition processes in development and learning* (pp. 31–58). New York: Cambridge University.

Stacey, K., & MacGregor, M. (1997). Ideas about symbolism that students bring to algebra. *Mathematics Teacher, 90*(2), 110–113.

Stagylidou, S., & Vosniadou, S. (2004). The development of students' understanding of the numerical value of fractions. *Learning and Instruction, 14*, 503–518.

Star, J. R. (2005). Reconceptualizing procedural knowledge. *Journal for Research in Mathematics Education, 36*(5), 404–411.

Stephens, A. C. (2005). Developing students' understandings of variable. *Mathematics Teaching in the Middle School, 11*(2), 96–100.

Swan, M. (2000). Making sense of algebra. *Mathematics Teaching, 171*, 16–19.

Sweller, J. (1999). *Instructional design in technical areas.* Camberwell, VIC, Australia: ACER Press.

Sweller, J., & Cooper, G. A. (1985). The use of worked examples as a substitute for problem solving in learning algebra. *Cognition and Instruction, 2*(1), 59–89.

Trafton, J. G., & Reiser, B. J. (1993). The contributions of studying examples and solving problems to skill acquisition. In M. Polson (Ed.), *Proceedings of the 15th annual conference of the cognitive science society* (pp. 1017–1022). Hillsdale, NJ: Erlbaum.

Tsamir, P., & Bazzini, L. (2004). Consistencies and inconsistencies in students' solution to algebraic 'single-value' inequalities. *International Journal of Mathematical Education in Science and Technology, 55*, 793–812.

Usiskin, Z. (1988). Conceptions of school algebra and uses of variables. In A. F. Coxford & A. P. Schulte (Eds.), *The ideas of algebra, K-12* (pp. 8–19). Reston, VA: National Council of Teachers of Mathematics.

Vaiyavutjamai, P., & Clements, M. A. (2006). Effects of classroom instruction on student performance on, understanding of, linear equations and linear inequalities. *Mathematical Thinking and Learning, 8*, 113–147.

Vamvakoussi, X., & Vosniadou, S. (2010). How many decimals are there between two fractions? Aspects of secondary school students' understanding of rational numbers and their notation. *Cognition and Instruction, 28*(2), 181–209.

Vamvakoussim, X., & Vosniadou, S. (2004). Understanding the structure of the set of rational numbers: A conceptual change approach. *Learning and Instruction, 14*(5), 453–467.

Van de Walle, J. A., Karp, K. S., & Bay-Williams, J. M. (2013). *Elementary and middle school mathematics: Teaching developmentally* (8th ed.). New York: Pearson Education.

Van Dooren, W., Lehtinen, E., & Vershcaffel, L. (2015). Unraveling the gap between natural and rational numbers. *Learning and Instruction, 37*, 1–4.

Van Dooren, W., Verschaffel, L., & Onghena, P. (2002). The impact of pre-service teachers' content knowledge on their evaluation of students' strategies for solving arithmetic and algebra word problems. *Journal for Research in Mathematics Education, 33*(5), 319–351.

Vlassis, J. (2002). About the flexibility of the minus sign in solving equations. In A. Cockburn & E. Nardi (Eds.), *Proceeding of the 26th conference for the international group of the psychology of mathematics education* (pp. 321–328). Norwich, UK: University of East Anglia.

Vlassis, J. (2004). Making sense of the minus sign or becoming flexible in 'negativity'. *Learning and Instruction, 14*, 469–484.

Vosniadou, S., & Brewer, W. F. (1992). Mental models of the earth: A study of conceptual change in childhood. *Cognitive Psychology, 24*(4), 535–585.

Warren, E. (2006). Comparative mathematical language in the elementary school: A longitudinal study. *Educational Studies in Mathematics, 62*, 169–189.

Watson, J. (1990). Research for teaching. Learning and teaching algebra. *Australian Mathematics Teacher, 46*(3), 12–14.

Wu, H. (2001). How to prepare students for algebra. *American Educator, 25*(2), 10–17.

Xin, Y. P., Wiles, B., & Lin, Y. (2008). Teaching conceptual model-based word problem story grammar to enhance mathematics problem solving. *The Journal of Special Education, 42*(3), 163–178.

Chapter 5
A Deep Understanding of Fractions Supports Student Success in Algebra

Stacy Reeder

Abstract Algebra is frequently referred to as the "gateway" course for high school mathematics in much the same way as calculus can "open" or "close" doors for students interested in pursuing degrees in science, technology, engineering, and mathematics (STEM) areas. This chapter presents the idea that students' challenges with algebra begin well before their first course in algebra and that these challenges are embedded in a complex set of issues. Weak or incomplete mathematical understanding of rational number concepts has a profound impact on students' success in algebra and subsequently, courses that follow where students are expected to confidently, competently, and efficiently address situations in which "and the rest is just algebra" is invoked. Recognizing that developing students' deep understanding of rational number concepts requires years of nurturing and care by capable, well-prepared teachers, both in terms of content and pedagogical knowledge, and a discussion of issues related to teacher preparation and teacher shortages and how these impact students' preparedness for algebra and their success in mathematics is presented.

Keywords Fractions • Algebra • Teacher education • Rational numbers • Proportional reasoning • Teacher shortages

Algebra is frequently referred to as the "gateway" course for high school mathematics in much the same way as calculus can "open" or "close" doors for students interested in pursuing degrees in science, technology, engineering, and mathematics (STEM) areas. This chapter presents the idea that students' challenges with algebra begin well before their first course in algebra and that these challenges are embedded in a complex set of issues. Weak or incomplete mathematical understanding of rational number concepts has a profound impact on students' success in algebra and subsequently, courses that follow where students are expected to confidently, competently, and efficiently address situations in which "and the

---

S. Reeder (✉)
University of Oklahoma, Stacy Reeder, 820 Van Vleet Oval, ECH 114,
Norman, OK 73019, USA
e-mail: reeder@ou.edu

S. Stewart (ed.), *And the Rest is Just Algebra*, DOI 10.1007/978-3-319-45053-7_5

rest is just algebra" is invoked. Recognizing that developing students' deep understanding of rational number concepts requires years of nurturing and care by capable, well-prepared teachers, both in terms of content and pedagogical knowledge, and a discussion of issues related to teacher preparation and teacher shortages and how these impact students' preparedness for algebra and their success in mathematics is presented.

5.1 Introduction

Over the last two decades, numerous reports have been written that focus on the need for improved mathematics and science teaching and learning in the United States. The pressure for global competitiveness and ever-changing demands of the workforce in the areas of science, technology, engineering, and mathematics (STEM) have propelled the conversation forward with intensity regarding learning outcomes in the STEM areas. In light of increased attention on STEM learning outcomes, the need for individuals prepared to enter the STEM fields, and, in general, the "need for more powerful learning focused on the demands of life, work, and citizenship in the twenty-first century" (Darling-Hammond, 2010), more students are taking algebra courses. The link to increased educational and economic opportunities has also been linked to the increase in the number of students taking algebra courses (Gamoran & Hannigan, 2000; Moses & Cobb, 2001; Nord et al., 2011; Rampey, Dion, & Donahue, 2009). Further, over the past several decades, and particularly since 2002 when the reauthorization of the Elementary and Secondary Education Act (ESEA) of 1965 commonly known as the "No Child Left Behind" Act attached passing exams based on algebra courses to graduation, more states require the passing of an algebra course for all students for graduation.

Algebra is frequently referred to as the "gateway" course for high school mathematics in much the same way as calculus can "open" or "close" doors for students interested in pursuing degrees in STEM areas. Stein, Kaufman, Sherman, and Hillen (2011) state that "[h]istorically, algebra has served a gatekeeper to advanced mathematics and science course taking and entry into high-paying, technical careers. Increased recognition of this phenomenon has led to a growing trend, . . . , for more students taking algebra in eighth grade" (p. 483). Their study examines algebra enrollment trends using data from the Early Childhood Longitudinal Study, Kindergarten class of 1988–99 (ECLS-K), the High School Transcript Study, National Assessment of Educational Progress (NAEP), NAEP Long-Term Trends, and Trends in International Mathematics and Science Study (TIMMS) and reveals a significant increase in algebra enrollment in eighth grade over the past two decades. Analysis of these sources provides empirical data that from the late 1980s to the early 1990s, enrollment in algebra for the nation's eighth graders had increased from 15 to 20 % to around 30 % in 2009. Additionally, their study reveals consistent lower enrolment in "eighth and ninth grade algebra among minorities and low-income students" (p. 460). This finding, along with an examination of policies related to who takes algebra and when students take algebra, called into question the preparedness of the students taking algebra. If the policies in place are universal, then it is likely many students taking algebra may not be prepared for the rigor and abstraction required for algebra. However, if the policies related to who takes algebra and when they take algebra allow for selection, evidence suggests that some prepared students from traditionally marginalized groups may be excluded from taking algebra prior to high school.

5.2 The Challenges of Algebra Preparedness

Research from various fields including mathematics education, mathematics teacher education, and mathematics reveals there is a confluence of issues that impact students' preparedness for algebra (e.g., Ball, 1993; Booth & Newton, 2012; Booth & Siegler, 2006, 2008; Harvey, 2012; Lamon, 2012; Ma, 1999; Newton, 2008; NMAP, 2008; Wu, 2001). Students' mathematical background and abilities, misconceptions and limitations related to their mathematical understanding, student self-confidence related to mathematics, policies related to the mathematics required in school prior to the taking of algebra, and teacher preparation for teaching mathematics at the elementary and middle school levels are among the chief contributors to this problem. In keeping with the title of this volume, "and the rest is just algebra," this chapter will present the argument that students' challenges with algebra begin well before their first course in algebra and that these challenges are embedded in a complex set of issues. While recognizing the complexity of this problem, this chapter will specifically explore the impact of weak or incomplete mathematical understanding of rational number concepts on students' success in algebra and subsequently, courses that follow. Also, included will be a discussion of issues related to teacher preparation and teacher shortages and how these impact students' preparedness for algebra and their success in mathematics.

5.3 Fraction Understanding Supports Algebra

The National Mathematics Advisory Panel (NMAP, 2008) suggests that a central goal of student's mathematical development is the conceptual understanding of fractions and procedural fluency with rational numbers and further implies that these competencies provide the critical foundation for algebra learning. Research corroborates the suggestions made by NMAP regarding the impact of weak or limited mathematical understanding at the elementary and middle school level and the significant impact it has on the future mathematical success of students and their educational possibilities (e.g., Booth & Newton, 2012; Wu, 2001). Brown and Quinn (2007) state that "vague fraction concepts and misunderstood fraction algorithms will ultimately be generalised into vague algebraic concepts and procedures. The lack of precise definitions and reliance upon shortcuts that are thoughtlessly given to students are likely to hinder performance in algebra" (p. 29).

Research has shown that much of the basis for algebraic understanding and algebraic thinking is contingent on a clear understanding of rational number concepts (Driscoll, 1982; Kieren, 1980; Lamon, 1999; Wu, 2001) and the ability to manipulate common fractions. For example, Booth and Newton (2012) found that "knowledge of fraction magnitudes—more so than whole number magnitude ... is related to students' skill in early algebra" (p. 251). Beyond simply using fractions and their related operations with fractions to solve algebraic problems involving fractions, students depend on their understanding of rates and ratios, often represented as fractions, to make sense of the key concepts of rate and variability in algebra. Wu (2001) claims that since operations with fractions can be generalized, fractions provide an opportunity to introduce students to the use of variables. Further, fractions are found throughout algebra. From coefficients to the slope of linear equations, from constants to solutions, from linear equations to completing the square, from solving systems of linear equations to solving rational equations, and from simple probabilities to the binomial theorem, algebra is brimming with examples that are directly and indirectly related to fractions. Wu (2001) suggests that "[w]ith proper infusion of precise definitions, clear explanations, and symbolic computations, the teaching of fractions can eventually hope to contribute to mathematics learning in general and the learning of algebra in particular" (p. 17).

Unfortunately, rational number concepts and fractions are challenging for many students, and students' understanding of rational numbers, or fractions, and misconceptions students might develop about fractions have a profound impact on their ability to learn algebra. According to Lamon (2012):

> Understanding fractions marks only the beginning of the journey toward rational number understanding. By the end of the middle school years, as a result of maturation, experience, and fraction instruction, it is assumed that students are capable of a formal thought process called proportional reasoning. This form of reasoning opens the door to high-school mathematics and science, and eventually, to careers in the mathematical sciences. The losses that occur because of the gaps in conceptual understanding about fractions, ratios, and related topics are incalculable. The consequences of doing, rather than understanding, directly or indirectly affect a person's attitudes towards mathematics, enjoyment and motivation in learning, course selection in mathematics and science, achievement, career flexibility, and even the ability to fully appreciate some of the simplest phenomena in everyday life (p. xi).

Algebra is replete with fractions and understanding many of the concepts found within algebra is dependent on student understanding of the multiple interpretations of fractions.

5.4 Deep Understanding of Fractions

Helping students develop a deep understanding and rich number sense about fractions and rational numbers including conceptual understanding and procedural fluency is not an easy task. It requires deep content knowledge specific to rational numbers on behalf of the teachers and requires several years to develop in students. Kieren (1988) reported that students in the United States rely heavily on rote memory of rules to solve fraction problems. The 2004 National Assessment of Educational Progress (NAEP), often referred to as the Nations Report Card, reported that 50 % of eighth grade students could not order three fractions from least to greatest and that fewer than 30 % of 17-year-olds correctly translated 0.029 as 29/1000 (Kloosterman, 2010). Further, Rittle-Johnson, Siegler, and Alibali (2001) conducted one-on-one controlled experiments and found that when asked which of two decimals 0.274 and 0.83 is greater, most fifth and sixth graders choose 0.274. Siegler et al. (2010) suggest that the lack of student conceptual understanding includes students not viewing fractions as numbers, viewing fractions as meaningless symbols that need to be manipulated in a variety of ways to produce answers that satisfy a teacher, focusing on numerators and denominators as separate numbers rather than thinking of the fraction as a single number, and confusing properties of fractions with those of whole numbers. They go on to state that "A high percentage of U.S. students lack conceptual understanding of fractions, even after studying fractions for several years; this, in turn, limits students' ability to solve problems with fractions and to learn and apply computational procedures involving fractions" (pp. 6–7).

The challenges are significant in the United States with regard to fraction and rational number teaching and student understanding. In light of these and other concerning findings, understanding this challenge and working to improve student learning related to fractions and rational numbers have been a focus of the mathematics education community for several decades. In the late 1980s, the publication of the National Council of Teachers of Mathematics (NCTM) Curriculum and Evaluation Standards (1989) and several other NCTM publications in the decade that followed helped drive the charge for change in fraction instruction. Since that time, there have been continual calls for fraction instruction to move from a procedural focus to one aimed at developing deep conceptual understanding (Lamon, 2012; Van de Walle, 2007). Understanding fractions concepts with depth is a complex endeavor and requires that teachers understand the work on fraction meanings and constructs. Kieren's work in the 1970s revealed the complexity of fraction understanding suggesting that the concept of fractions consists of several sub-constructs or meanings (1976).

In his work, Kieren suggested that one must understand each sub-construct independently and jointly in order to have a general understanding of fractions. Initially, Kieren identified four meanings for fractions: measure, ratio, quotient, and operator. Originally, the notion of the part-whole relationship served as a basis for the development of the other sub-constructs and as such was not included in the list

Interpretations of $\frac{4}{5}$	Meaning
Part-Whole Comparisons with Unitizing "4 parts out of 5 equal parts"	$\frac{4}{5}$ means four parts out of five equal parts of the unit, with equivalent fractions found by thinking of the parts in terms of larger or smaller chunks.
Measure "4 ($\frac{1}{5}$ – units)"	$\frac{4}{5}$ means a distance of 4 ($\frac{1}{5}$ – units) from 0 on the number line or 4 ($\frac{1}{5}$ – units) of a given area.
Operator "$\frac{4}{5}$ of something"	$\frac{4}{5}$ gives a rule that tells how to operate on a unit (or on the result of a previous operation); multiply 4 and divide your result by 5 or divide by 5 and multiply the result by 4. This results in multiple meanings for $\frac{4}{5}$; 4 ($\frac{1}{5}$ – units), 1 ($\frac{4}{5}$ – units), and $\frac{1}{5}$ (4 – units).
Quotient "4 divided by 5"	$\frac{4}{5}$ is the amount each person receives when 5 people share a 4 – unit of something.
Ratios "4 to 5"	4:5 is a relationship in which there are 4 A's compared, in a multiplicative rather than an additive sense, to 5 B's.

Fig. 5.1 Fraction interpretations and meanings (adapted from Lamon, 2012)

as a separate construct. Kieren's ideas were later expanded by Behr, Lesh, Post, and Silver (1983) who recommended that the part-whole relationship as seen by Kieren to be embedded in the four other meanings be considered a distinct sub-construct of fractions (see Fig. 5.1). Their work connected the part-whole meaning of fractions with the notion of portioning and establishing it as a distinct sub-construct of fractions. Behr et al.'s (1983) work revealed that the process of partitioning and the part-whole sub-construct of rational numbers are fundamental for developing a deep understanding of the four other constructs of fractions. Since that time, others (Lamon, 1999, 2012; Mack, 2001; Simon, 1993; Tobias, 2012) have suggested that conceptualizing the whole is important for understanding many significant mathematical concepts including contextualizing situations, understanding procedures, and interpreting solutions.

The notion of part-whole as a construct for fractions and rational numbers occupies a significant place in curricular materials for elementary children throughout the world. This is based on the assumption that conceptualizing the whole and understanding part-whole relationships is fundamental to many important mathematical concepts including the four constructs of fractions identified by Kieren (1976) and that operations with fractions are connected to the part-whole relationship (Behr et al., 1983). Lamon (2012) discusses the idea that more emphasis should not be placed on one sub-construct, or interpretation, of fractions and that rather, teachers should understand that no single interpretation is a panacea. Cramer and Whitney (2010), however, suggested that the part-whole sub-construct is a good place for children to begin to develop an understanding of fractions.

Many researchers agree but also believe that while the part-whole meaning of fractions is the most commonly relied upon interpretation in curricular materials, placing more emphasis on other interpretations would help students gain a better understanding of fractions (e.g., Clarke, Roche, & Mitchell, 2008; Siebert & Gaskin, 2006). This suggests that an emphasis on the part-whole construct or interpretation of fractions, while perhaps serving as a basis for understanding fractions, is not sufficient by itself for deep understanding and flexibility with fractions. Lamon (2007) believes that we have a tremendous problem related to fraction teaching and learning due to the fact that most teachers only understand and teach fractions from a part-whole understanding. The findings of a study conducted by Reeder and Utley (under review) focused on prospective elementary teachers corroborates this claim. The prospective elementary teachers in their study relied almost exclusively on part-whole understanding of fractions as part of a whole to answer basic questions about fractions, and when asked how they would explain the concept of fractions to their students, the majority of the participants provided a part-whole explanation.

5.5 The Importance of Proportional Reasoning for Algebra

While there are functional differences between each of the five sub-constructs of fractions, they are interrelated. In addition, it is believed that, fractions should be taught in such a way that students develop a holistic understanding of fractions that includes the multiple perspectives of each of the sub-constructs. In this way, students may be able to work more flexibly within varied contexts, with more representations, and develop the higher-order thinking needed for proportional reasoning (Lesh, Post, & Behr, 1988). However, the sub-construct of ratio and rates is most related to proportional reasoning which makes it of paramount importance for student success in algebra. Proportional reasoning has been referred to as the cornerstone of higher levels of mathematics success (Kilpatrick, Swafford, & Findell, 2001; Lamon, 1999; Lesh et al., 1988). Wright (2005) states that proportional reasoning involves "making multiplicative comparisons between quantities" (p. 363), and Lesh et al. (1988) add that it is "the ability to mentally store and process several pieces of information" (p. 93). According to Lamon (1999), "proportional reasoning is one of the best indicators that a student has attained understanding of rational numbers" (p. 3).

The ability to reason proportionally involves a student's ability to understand variation and covariation and make multiple comparisons. It involves students' abilities to differentiate between relative and absolute meanings of "more" and determine which of these is a proportional relationship, compare ratios without using common denominator algorithms, differentiate between additive and multiplicative processes and their effects on scale and proportionality, and interpret graphs that represent proportional relationships or direct and indirect variation. These abilities are directly related to the kind of thinking and reasoning needed for algebraic reasoning and developing an understanding of functions. For example, Lobato and Thanheiser (2002) discuss the need for students to understand slope "as a ratio that measures some attribute in a situation" (p. 174). They go on to argue that helping students understand "the modeling and proportional reasoning aspects of ratio-as-measure tasks, can in turn help students develop an understanding of slope that is more general and applicable" (p. 174) and important for success in algebra.

5.6 Challenges Regarding Preparation of Teachers of Mathematics

Research in mathematics education has also well documented the challenges of teaching rational number concepts and the impact of teachers' limited content knowledge on their students' learning (e.g., Ball, 1993; Harvey, 2012; Lamon, 2012; Ma, 1999; Newton, 2008). The Conference Board of the Mathematical Sciences (CBMS, 2012) states that "a critical pillar of a strong PreK–12 education is a well-qualified teacher in every classroom" (p. 14). Unfortunately, that is not always the case with regard to teachers of mathematics at all grade levels. The paths to teacher certification in the United States are varied allowing for significant difference in what and how much mathematics is required for credentialing. Many states, due to a decade's long shortage of mathematics teachers, allow individuals prepared to teach elementary, many of whom have had little college level mathematics, to simply pass an exam to receive credentials to teach middle level mathematics—in some cases up through Algebra II. Sadly, with these extreme teacher shortages across the nation, some states are allowing significant numbers of individuals into mathematics classrooms with little or no background in mathematics.

In the case of teachers who have completed a teacher preparation program, the challenges and limitations related to their content knowledge for teaching have been a focus of the mathematics education community for decades and have been well documented in the mathematics education literature (Ball, 1993; CBMS, 2001, 2012; Ma, 1999; Shulman, 1986). For education practice, policy, and research, teachers' mathematical content knowledge continues to be a major focus (CBMS, 2012; Greenberg & Walsh, 2008; National Mathematics Advisory Panel, 2008). Despite this ongoing focus, a great number of teachers, particularly those teaching in elementary, intermediate, and middle level mathematics, continue to be under-prepared and uncomfortable with the mathematics content they are expected to teach (Greenberg & Walsh, 2008). This is often due to a variety of factors including, but not limited to, their own experiences with mathematics, their beliefs and ideas about mathematics teaching and learning, and their preparation as teachers related to mathematics content knowledge and pedagogical knowledge for teaching mathematics (Reeder, Utley, & Cassel, 2009; Utley & Reeder, 2012).

Prior to their teacher preparation coursework, most prospective teachers have spent many years learning mathematics from teachers whose pedagogical practices primarily reflect a traditional orientation focused on procedural understanding rather than a balanced approach that attends to both conceptual understanding and procedural fluency (National Research Council, 2001). When they arrive in their undergraduate degree, they are typically engaged with mathematics similar to their prior experiences through lecture style teaching methods and a show-and-repeat procedures approach. Further, many teacher preparation programs require prospective teachers to take mathematics coursework that is disconnected from the mathematics they will teach. Prospective mathematics teachers are required in many states to take a course in College Algebra which may extend their own mathematical understanding but does not do much to deepen their understanding of rational numbers, for example. This certainly shapes teachers' attitudes about mathematics and their ideas about what constitutes mathematics teaching and learning (Reeder et al., 2009). Likewise, most secondary mathematics education programs preparing teachers to teach grades 6–12 mathematics require, if not a degree in applied mathematics, the coursework equivalent. Prospective secondary mathematics teachers are typically required to take coursework well beyond what many consider as necessary the strong content knowledge needed for teaching but very well may not understand rational number concepts with depth. The CBMS recommends more mathematics coursework specifically developed to meet the needs of teachers and improve content knowledge specifically needed for teaching (2012).

Specific to this chapter, existing research demonstrates that prospective and in-service teachers' knowledge of fractions is limited (Ball, 1990; Becker & Lin, 2005; Chinnappan & Forrester, 2014; Cramer, Post, & del Mas, 2002; Harvey, 2012; Ma, 1999; Newton, 2008; Zhou, Peverly, & Xin, 2006). Additionally, research has documented that teaching and learning fraction concepts are a difficult and complex undertaking (Ball, 1993; Harvey, 2012; Lamon, 2012; Ma, 1999; Newton, 2008). Newstead and Murray (1998) purport that fractions are among the most complex mathematical concepts that elementary students encounter, and Charalambous and Pitta-Pantazi (2005) and Harvey (2012) assert that the teaching and learning of fractions have traditionally been problematic. Lamon (2007) believes that most teachers are not prepared to teach content other than the part-whole construct of fractions which leaves their students with an incomplete and shallow understanding of fractions and rational numbers.

5.7 The Growing Problem of Teacher Shortages

The United States is in the midst of a teacher shortage crisis. For decades there has been a chronic shortage in particular teaching content areas such as mathematics, science, special education, and bilingual education, but the current situation is widespread and involves almost every state in the nation. From California to

Oklahoma to New York, school districts are scrambling to hire teachers, and unfortunately, in many cases this is regardless of their credentialing. In October 2015, US World and News Report reported that school districts in the state of California were still trying to fill 21,500 vacant teaching positions. In this same report, Partelow stated that "while it may be too early to tell whether this year's reported shortages are a blip or part of a long-term systemic trend, we do know that fewer college students are enrolling in teacher training programs and surges of teachers are retiring" (2015, para. 4). If this trend continues, it will not only lead to greater numbers of unfilled teaching positions in the future but will also lead to classrooms likely filled with teachers who are not as well prepared as needed.

When the school year begins each fall and there are not enough teachers to fill the classrooms in each building, students do not sit in empty rooms. Rather, school districts begin filling classrooms, in some cases, with anyone they can find regardless of the person's credentials. This results in credentialed teachers teaching outside of their content area, long-term substitutes filling teaching positions, preservice teachers beginning teaching before they are fully prepared, and allowances for individuals to be "emergency certified" often without any teacher preparation. California, for example, has been particularly hard-hit following the loss of more than 80,000 teaching jobs between 2008 and 2012 (Rich, 2015). Now, with a recovering economy, there is a need for more teachers and they simply are not enough. Rich (2015) reported that "[b]efore taking over a classroom solo in California, a candidate typically must complete a post-baccalaureate credentialing program, including stints as a supervised student teacher. But in 2013–2014, the last year for which figures are available, nearly a quarter of all new teaching credentials issued in California were for internships that allowed candidates to work full time as teachers while simultaneously enrolling in training courses at night or on weekends" (para. 13). Additionally, from 2012 to 2013, the number of emergency permits issued in California to allow individuals who have no teaching credentials to fill teaching positions jumped by more than 36 %. This increase has been unfortunately paralleled in other states in the past few years. Partelow (2015), citing Oklahoma as an example, stated that "[u]nfortunately some states have instead responded [to the teacher shortage] by lowering the (arguably too low already) bar for entry into the profession. Oklahoma approved over 800 emergency certificates in July and August allowing non-credentialed teachers to teach in classrooms of their own" (para. 7). In October of the fall 2015 semester, over 1000 teacher vacancies remained unfilled in Oklahoma.

The teacher shortage will undoubtedly have an impact on students' mathematical preparedness. More classrooms will be filled with teachers who do not have the specialized content knowledge needed for teaching mathematics or the pedagogical content knowledge to teach mathematics effectively. Without deep content knowledge or sophisticated and well-developed pedagogical practices, teachers typically resort to teaching via rote methods and memorization—methods that do not account for a holistic approach to teaching fractions and rational number concepts with the five sub-constructs in mind.

5.8 Discussion

In a recent report published by the National Academy of Sciences, the author stated that the phrase "STEM education is shorthand for an enterprise that is as complicated as it is important" (Beatty, 2011, p. 1). She goes onto to say that:

> what students learn about the science disciplines, technology, engineering, and mathematics during their K-12 schooling shapes their intellectual development, opportunities for future study and work, and choices of career, as well as their capacity to make informed decisions about political and civic issues and about their own lives. A wide array of public and personal issues—from global warming to medical treatment to social networking to home mortgages—involve science, technology, engineering, and mathematics (STEM). Indeed, the solutions to some of the most daunting problems facing the national will require not only the expertise of top STEM professionals but also the wisdom and understanding of its citizens. (Beatty, 2011, p. 1)

Clearly, helping students succeed in STEM fields and to live and succeed in a global economy is important, and simply engaging students in the mastery of basic skills is not sufficient to meet this goal.

In his popular book, *The Checklist Manifesto*, Atul Gawande (2010) addresses the idea that despite our modern world and tremendous advances in health care, government, the law, and financial industry, challenges still plague us. He examines the nature of problems we frequently face and elaborates on the nature and complexity of said problems. Referencing the work of Glouberman and Zimmerman (2002), Gawande presents three different kinds of problems in the world: the simple, the complicated, and the complex. Simple problems, he notes, "are ones like baking a cake from a mix. There is a recipe and a few basic techniques to follow but once these are mastered, following the recipe brings a high likelihood of success" (p. 49). Complicated problems on the other hand, are ones like sending a rocket to the moon. "They can sometimes be broken down into a series of simple problems but there is no straightforward recipe. Success frequently requires multiple people, often multiple teams, and specialized expertise" (Gawande, 2010, p. 49), but once you learn to send a rocket to the moon, you can repeat the process with other rockets and perfect it—one rocket is typically like another rocket. "Complex problems, however, are like raising a child. Although raising one child may provide experience, it does not guarantee success with the next child" (Gawande, 2010, p. 49). Expertise is valuable but likely not sufficient because unlike rockets, every child is unique. Each child may require an entirely different approach from the previous one. Another feature of complex problems is that their outcomes remain highly uncertain. "Yet we all know that it is possible to raise a child well. It's complex, that's all" (Gawande, 2010, p. 49). Likewise, helping students be prepared for algebra is a complex endeavor.

Preparing students well for algebra involves many years of working with them to develop a deep understanding of fractions and proportional reasoning and ensuring that our teachers not only understand but are able to teach rational number concepts holistically. The challenge is multifaceted involving policy and practice, beliefs about mathematics teaching and learning, beliefs about what is mathematics and what it means to know mathematics deeply, teacher preparation, and ensuring every classroom of students has a well-qualified teacher. The complex challenge of helping students be prepared for algebra and the important mathematics beyond algebra involve many years of work and development. Equally important is the specialized content and the pedagogical knowledge of many skillful teachers who teach mathematics.

Darling-Hammond (2010) states that we can meet the challenges of our current education system by developing a new paradigm for national and state education policy that is guided "by twin commitments to *support meaningful learning* on the part of students, teachers, and schools and to *equalize access to educational opportunity*, making it possible for all students to profit from more productive schools" (p. 278). If, as an education community, we believe in the importance of preparing students to live happily and succeed in a global economy, then we need to insist that the mastery of basic skills that the emphasis on accountability has brought is not sufficient. Cortese and Ravitch (2008) noted that "[W]hat we need is an education system that focuses on deep knowledge, that values creativity and originality, and that values thinking skills" (p. 4). As an education community, we can advocate for policies and practices that support the teaching of deep knowledge and support teachers in helping students learn meaningfully. These challenges can be met and when they are, we can be confident that students in calculus courses and beyond will respond competently and efficiently when addressing "and the rest is just algebra."

References

Ball, D. L. (1990). Breaking with experience in learning to teach mathematics: The role of a preservice methods course. *For the Learning of Mathematics, 10*(2), 10–16.

Ball, D. L. (1993). Halves, pieces, and twoths: Constructing and using representational contexts in teaching fractions. In *Rational numbers: an integration of research* (pp. 157–195).

Beatty, A. (Ed.). (2011). *Successful STEM education: A workshop summary*. Washington, DC: National Academies Press.

Becker, J. P. & Lin, C.Y. (2005). Effects of a computational skills workshop on prospective elementary teachers. Preliminary report. In *Paper presented at the annual meeting of the MAA and the American mathematical society, Atlanta, GA*.

Behr, M., Lesh, R., Post, T., & Silver, E. (1983). Rational number concepts. In R. Lesh & M. Landau (Eds.), *Acquisition of mathematics concepts and processes* (pp. 91–125). New York: Academic.

Booth, J. L., & Newton, K. J. (2012). Fractions: Could they really be the gatekeeper's doorman? *Contemporary Educational Psychology, 37*(4), 247–253.

Booth, J. L., & Siegler, R. S. (2006). Developmental and individual differences in numerical estimation. *Developmental Psychology, 41*, 189–201.

Booth, J. L., & Siegler, R. S. (2008). Numerical magnitude representations influence arithmetic learning. *Child Development, 79*(4), 1016–1031.

Brown, G., & Quinn, R. J. (2007). Fraction proficiency and success in algebra: What does research say? *Australian Mathematics Teacher, 63*(3), 23–30.

Charalambous, C. Y., & Pitta-Pantazi, D. (2005). Revisiting a theoretical model on fractions: Implications for teaching and research. *International group for the psychology of mathematics education* (p. 233).

Chinnappan, M., & Forrester, T. (2014). Generating procedural and conceptual knowledge of fractions by pre-service teachers. *Mathematics Education Research Journal, 26*(4), 871–896.

Clarke, D. M., Roche, A., & Mitchell, A. (2008). 10 tips for making fractions come alive and make sense. *Mathematics Teaching in the Middle School, 13*(7), 372–380.

Conference Board of the Mathematical Sciences. (2001). *The mathematical education of teachers*. Washington, DC: American Mathematical Society and Mathematical Association of America.

Conference Board of the Mathematical Sciences. (2012). *The mathematical education of teachers II*. Washington, DC: American Mathematical Society and Mathematical Association of America.

Cortese, A. & Ravitch, D. (2008). Still at risk: What students don't know, even now. In: *A preface in the report from common core by Frederick M. Hess*. Washington, DC: Common Core.

Cramer, K., Post, T. R., & del Mas, R. C. (2002). Initial fraction learning by fourth and fifth grade students: A comparison of the effects of using commercial curricula with the effects of using the rational number project curriculum. *Journal for Research in Mathematics Education, 33*, 111–144.

Cramer, K., & Whitney, S. (2010). Learning rational number concepts and skills in elementary school classrooms. In D. V. Lambdin & F. K. Lester Jr. (Eds.), *Teaching and learning mathematics: Translating research for elementary school teachers* (pp. 15–22). Reston, VA: National Council of Teachers of Mathematics.

Darling-Hammond, L. (2010). *The flat world and education: How America's commitment to equity will determine our future*. New York: Teachers College Press.

Driscoll, M. (1982). Research within reach: Secondary school mathematics. *A research guided response to the concerns of educators. Missouri*: (ERIC Document Reproduction Service No. ED 225 842).

Gamoran, A., & Hannigan, E. C. (2000). Algebra for everyone? Benefits of college-preparatory mathematics for students with diverse abilities in early secondary school. *Educational Evaluation and Policy Analysis, 22*(3), 241–254.

Gawande, A. (2010). *The checklist manifesto: How to get things right*. New York: Metropolitan Books.

Glouberman, S., & Zimmerman, B. (2002). *Complicated and complex systems: What would successful reform of Medicare look like?* Toronto: Commission on the Future of Health Care in Canada.

Greenberg, J., & Walsh, K. (2008). *No common denominator: The preparation of elementary teachers in mathematics by America's education schools*. National Council on Teacher Quality.

Harvey, R. (2012). Stretching student teachers' understanding of fractions. *Mathematics Education Research Journal, 24*, 493–511.

Kieren, T. E. (1976). On the mathematical, cognitive and instructional. In *Number and measurement. Papers from a research workshop* (Vol. 7418491, p. 101).

Kieren, T. E. (1980). The rational number construct: Its elements and mechanisms. In T. E. Kieren (Ed.), *Recent research on number learning* (pp. 125–149). Columbus: Ohio State University (ERIC Document Reproduction Service No. ED 212 463).

Kieren, T. E. (1988). Personal knowledge of rational numbers: Its intuitive and formal development. In J. Hiebert & M. Behr (Eds.), *Number concepts and operations in the middle grades* (pp. 162–181). Reston, VA: National Council of Teachers of Mathematics.

Kilpatrick, J., Swafford, J., & Findell, B. (2001). *Adding it up: Helping children learn mathematics*. Washington, DC: National Academy Press.

Kloosterman, P. (2010). Mathematics skills of 17-year-olds in the United States: 1978 to 2004. *Journal for Research in Mathematics Education, 41*(1), 20–51.

Lamon, S. J. (1999). *Teaching fractions and ratios for understanding: Essential content knowledge and instructional strategies for teachers*. Mahwah, NJ: Lawrence Erlbaum.

Lamon, S. J. (2007). Rational numbers and proportional reasoning: Toward a theoretical framework for research. *Second Handbook of Research on Mathematics Teaching and Learning, 1*, 629–667.

Lamon, S. J. (2012). *Teaching fractions and ratios for understanding: Essential content knowledge and instructional strategies for teachers* (3rd ed.). Mahwah, NJ: Lawrence Erlbaum Associates.

Lesh, R., Post, T., & Behr, M. (1988). Proportional reasoning. *Number Concepts and Operations in the Middle Grades, 2*, 93–118.

Lobato, J., & Thanheiser, E. (2002). Developing understanding of ratio-as-measures a foundation for slope. In B. Litwiller & G. Bright (Eds.), *Making sense of fractions, ratios, and proportions*. Reston, VA: National Council of Teachers of Mathematics.

Ma, L. (1999). *Knowing and teaching elementary mathematics: Teachers' understanding of fundamental mathematics in China and the United States*. Mahwah, NJ: Lawrence Erlbaum Associates.

Mack, N. K. (2001). Building on informal knowledge through instruction in a complex content domain: Partitioning, units, and understanding multiplication of fractions. *Journal for Research in Mathematics Education, 32*(3), 267–295. doi:10.2307/749828.

Moses, R. P., & Cobb, C., Jr. (2001). Organizing algebra: The need to voice a demand. *Social Policy, 4*, 12.

National Council of Teachers of Mathematics. (1989). *Curriculum and evaluation standards for school mathematics*. Reston, VA: Author.

National Mathematics Advisory Panel. (2008). *Foundations for success: The final report of the national mathematics advisory panel*. Washington, DC: US Department of Education.

National Research Council. (2001). *Adding it up: Helping children learn mathematics*. Washington, DC: National Academies Press.

Newstead, K., & Murray, H. (1998). Young students' constructions of fractions. In A. Olivier & K. Newstead (Eds.), *Proceedings of the twenty-second international conference for the psychology of mathematics education* (Vol. 3, pp. 295–302). South Africa: Stellenbosch.

Newton, K. J. (2008). An extensive analysis of preservice elementary teachers' knowledge of fractions. *American Educational Research Journal, 45*(4), 1080–1110.

Nord, C., Roey, S., Perkins, R., Lyons, M., Lemanski, N., Brown, J., et al. (2011). *The nation's report card: America's high school graduates (NCES 2011–462)*. Washington, DC: U.S. Department of Education, National Center for Education Statistics.

Partelow, L. (2015). You're hired (to teach, for the right reasons). *US News and World Report*. Retrieved from http://www.usnews.com/opinion/knowledge-bank/2015/10/14/address-americas-teacher-shortage-the-right-way.

Rampey, B. D., Dion, G. S., & Donahue, P. L. (2009). NAEP 2008: Trends in academic progress. NCES 2009-479. *National Center for Education Statistics*.

Reeder, S. & Utley, J. (Under review). *What is a fraction? Developing fraction understanding in prospective elementary teachers*.

Reeder, S., Utley, J., & Cassel, D. (2009). Using metaphors as a tool for examining preservice elementary teachers' beliefs about mathematics teaching and learning. *School Science and Mathematics, 109*(5), 290–297.

Rich, M. (2015). Teacher shortages spur a nationwide hiring scramble (credentials optional). *New York Times*. Retrieved from http://www.nytimes.com/2015/08/10/us/teacher-shortages-spur-a-nationwide-hiring-scramble-credentials-optional.html?_r=0.

Rittle-Johnson, B., Siegler, R. S., & Alibali, M. W. (2001). Developing conceptual understanding and procedural skill in mathematics: An iterative process. *Journal of Educational Psychology, 93*(2), 346–362.

Shulman, L. S. (1986). Those who understand: Knowledge growth in teaching. *Educational Researcher, 15*(2), 4–14.

Siebert, D., & Gaskin, N. (2006). Creating, naming, and justifying fractions. *Teaching Children Mathematics, 12*(8), 394–400.

Siegler, R., Carpenter, T., Fennell, F., Geary, D., Lewis, J., Okamoto, Y, Thompson, L & Wray, J. (2010). *Developing effective fractions instruction for Kindergarten through 8th grade. IES practice guide*. NCEE 2010-4039. What Works Clearinghouse.

Simon, M. A. (1993). Prospective elementary teachers' knowledge of division. *Journal for Research in Mathematics education*, 233–254.

Stein, M. K., Kaufman, J. H., Sherman, M., & Hillen, A. F. (2011). Algebra a challenge at the crossroads of policy and practice. *Review of Educational Research, 81*(4), 453–492.

Tobias, J. M. (2013). Prospective elementary teachers' development of fraction language for defining the whole. *Journal of Mathematics Teacher Education, 16*, 85–103. doi:10.1007/s10857-012-9212-5.

Utley, J., & Reeder, S. (2012). Prospective elementary teachers development of fraction number sense. *Investigations in Mathematics Learning, 5*(2), 1–13.

Van de Walle, J. A. (2007). *Elementary and middle school mathematics: Teaching developmentally* (6th ed.). Boston, MA: Pearson.

Wu, H. (2001). How to prepare students for algebra. *American Educator, 25*(2), 10–17.

Zhou, Z., Peverly, S. T., & Xin, T. (2006). Knowing and teaching fractions: A cross-cultural study of American and Chinese mathematics teachers. *Contemporary Educational Psychology, 31*(4), 438–457.

Part III
Positive Approaches to the Teaching of Algebra

Part III
Positive Approaches to the Teaching of Algebra

Chapter 6
Overcoming the Algebra Barrier: Being Particular About the General, and Generally Looking Beyond the Particular, in Homage to Mary Boole

John Mason

Algebra consists in preserving a constant, reverent, and conscientious awareness of our own ignorance [p. 56]
Teaching involves preventing mechanicalness from reaching a degree fatal to progress [p. 15]
The use of algebra is to free people from bondage [p. 56]
[all quotes are from Mary Boole, extracted in Tahta, 1972]

Abstract Consistent with a phenomenographic approach valuing lived experience as the basis for future actions, a collection of pedagogic strategies for introducing and developing algebraic thinking are exemplified and described. They are drawn from experience over many years working with students of all ages, teachers and other colleagues, and reading algebra texts from the fifteenth century to the present. Attention in this chapter is mainly focused on invoking learners' powers to express generality, to instantiate generalities in particular cases, and to treat all generalities as conjectures which need to be justified. Learning to manipulate algebra is actually straightforward once you have begun to appreciate where algebraic expressions come from.

Keywords Expressing generality • Pedagogic strategies • Tracking arithmetic • Watch What You Do • Say What You See • Reasoning without numbers • Same and different • Invariance in the midst of change

J. Mason (✉)
27 Elms Rd, Oxford OX2 9JZ, UK
e-mail: John.Mason@open.ac.uk

S. Stewart (ed.), *And the Rest is Just Algebra*, DOI 10.1007/978-3-319-45053-7_6

6.1 Introduction

That algebra is a watershed for most learners is common experience, and it has been the case ever since algebra emerged. It has long been my claim that school algebra is fundamentally the expression of generality in a succinct form so that it can be manipulated (Mason, Graham, Pimm, & Gowar, 1985). The fact that almost all books on algebra (or arithmetic with algebra) since the fifteenth century have introduced algebra as *the manipulation of letters as if they were numbers* suggests that recognition of algebra as expression of generality seems so obvious as not to require mentioning, while what teachers want students to achieve is facility in manipulating algebraic expressions. Consequently the usual focus is on how to manipulate algebraic expressions. Or it could be that the constant pressure to get learners to perform, to carry out procedures, has blinded curriculum designers to the essence of algebra.

It seems to have been Isaac Newton (1683) who diverted attention from the expression of generality to the nuts and bolts of algebraic manipulation, namely the solving of equations, though some of his contemporaries questioned whether expressing generality was as straightforward and simple as he claimed (Ward, 1706). Pushing learners immediately into solving equations (first linear, then quadratic then perhaps factored or factorable polynomials and perhaps then into iterative methods for approximate solutions) is a reflection of the technician's approach, the result of a particular *transposition didactique* (Chevallard, 1985): on discovering a formula or a method, students are then faced with that method, usually without the insight that led to it. But why would learners want to internalise a collection of procedures involving entities that have no meaning for them? My claim has always been that unless learners appreciate where equations come from, unless they comprehend the origins of equations and inequalities in the expression of generality, algebraic expressions and algebra itself will remain a mystery, and a watershed.

That algebra as the manipulation of letters is mysterious has been attested to by generations of learners concerning their experience at school. Many claim that they could do what was asked, but had no idea what it was about or why they were doing it. Recent generations have become less willing to undertake what seems to them meaningless, resulting in algebra continuing to be one of the major watersheds of school mathematics.

Yet there is abundant evidence that young children can cope with abstraction, even with symbols for the as-yet-unspecified. Weakness in algebraic manipulation comes, I claim, not from insufficient practice, but from teachers concentrating on manipulation rather than invoking and evoking learners' natural powers to specialise and to generalise, to see the general through the particular and to see the particular in the general (Mason & Pimm, 1984).

6.2 Methods

I am interested in what is possible, happy that others are concerned to study what is the case currently in their situation. Furthermore, I am interested in lived experience, and as such I am committed to taking a phenomenological stance. Thus in this chapter the reader will find numerous mathematical tasks through and by means of which it is possible to get a taste of the more general claims that I am making. I am convinced that this is the best way to work with learners and colleagues: to offer experiences which can form the basis for noticing what might previously have passed by unnoticed, thereby sensitising oneself to notice opportunities to support and promote others becoming aware of something similar for themselves. This has been the basis for Open University courses for teachers since 1982 (Mason, Graham, Pimm, & Gowar 1985; Open University, 1982), and a foundation for research as elaborated in *Researching Your Own Practice*: *the discipline of noticing* (Mason, 2002a).

I offer no programme, no recommended or researchable sequence of tasks that will prove to be most effective. Rather my approach is to work on developing sensitivities to possibilities so that potential actions come to mind in the moment (actually, *come to action* but are consciously considered before being enacted) when they are needed. Thus the teacher can be attending to what learners are saying and doing, rather than to a prepared sequence of tasks. This is in line with the notion of *teaching by listening* (Davis, 1996).

6.3 Being Particular About the General

The suggestion in this section is that being particular about invoking and evoking generality, placing the expression of generality at the heart of the curriculum (and not simply in mathematics) would benefit many learners who for some reason or other, seem to leave their natural powers at the classroom door. There is extensive research backing up this proposition stretching over many years. See, for example, Giménez, Lins, and Gómez (1996), Bednarz, Kieran, and Lee (1996), Chick, Stacey, Vincent, and Vincent (2001), Mason and Sutherland (2002), Kaput, Carraher, and Blanton (2008) and Cai and Knuth (2011).

6.3.1 Beginning in the Earliest Years

Mary Boole finds the origins of algebra in young children's experience such as that a metal teapot can be hot or cold: some of its attributes can vary (Tahta, 1972, pp. 57–58). Notice that there is an inherent use of what has come to be called *variation theory* which suggests that what is available to be learned is what has been experienced as varying in close proximity of time and space (Marton, 2015; Marton & Booth, 1997). Even earlier in a child's life, in order to recognise mother in her various guises, with different smells and appearances, it is necessary to generalise, to recognise that some attributes can change while others remain invariant. This applies in the affective-emotional domain just as it does in the physical-enactive domain, and in the cognitive-intellectual domain. Indeed, as Caleb Gattegno (1988) claimed, the foetus in the womb already shows signs of generalising, responding to different stimuli in particular ways.

To learn to read people's expressions, to learn to grab and put things in your mouth, to crawl, to stand, to walk and to talk all require extensive and wide-ranging use of natural powers to specialise and generalise. It has often been said that, given our success in teaching children to read and write, it is a good thing we don't have to teach children to talk as well. Put another way, having used and developed their natural powers so well before they reach school, how might we call upon those same powers to develop further, so that reading and writing, counting and arithmetic, algebra and conceptual thinking are just as natural? Terezinha Nunes and Peter Bryant (1996) (see also Nunes, Bryant, & Watson, 2008) show clearly how making use of what children bring to school in the way of experience and internalised actions can make a substantial difference to the children's experience and success in school.

Western approaches have been strongly influenced by the staircase metaphor for learning, in which learners gradually ascend a staircase of 'levels' from the simple to the complex, from the particular to the more general, from the specific to the abstract. This permeates both curriculum and pedagogy. Jerome Bruner (1966) distinguished three modes of (re)presentation (enactive, iconic and symbolic). Considered by researchers, curriculum designers, mathematics educators, and teachers as a sequence rather than as three worlds of experience between which we move as we add layers of appreciation, comprehension and hence understanding, learners have often been enculturated into a sequence of always building from the simple to the complex, the particular towards the general, the concrete towards the abstract. Because this is how we teach, many learners balk at some stage and so do not experience the general, the abstract, the overview. They remain locked into the specifics of procedures without appreciation of what is possible, without comprehension of what can be achieved, and without understanding of what their actions are all about. Mary Boole warned against this, but generations of learners are still having the experience of 'hopeless non-comprehension', or even of 'self-protecting and contemptuous non-attention' (Tahta, 1972, p. 51). She recommended 'build[ing] up good habits on a basis within which falls the centre of gravity of the individual with whom you are dealing with' (Tahta, 1972, p. 17).

A contrasting approach has been promoted by Vasily Davydov (1990) and taken up by Jean Schmittau (2004) and Barbara Dougherty (2008), among others, who have shown that young children are perfectly capable of working from abstractions and generality to instantiation in particular situations.

An intermediate stance is both possible and desirable: sometimes starting from particulars, sometimes from a slight or moderate generality and sometimes from an extremely general statement. Learners are then encouraged, whenever they are stuck, to specialise to examples with which they are more confident, and then to re-generalise as they begin to make sense of underlying structure. The purpose of specialising is not to fill a notebook with examples, but rather to detect and try to express underlying structural relationships.

This process was summarised as a pedagogic strategy and as a learning strategy in Open University (1982); see also Mason, 2002b or Mason & Johnston-Wilder, 2004) as a continuing spiral of *Manipulating – Getting a sense of—Articulating—Manipulating—Getting a sense of – Articulating—*. . . . This means turning to confidence-inspiring entities, manipulating them in order to locate structural relationships, getting a sense of what is going on, and trying to articulate this, eventually reaching a succinct articulation which can form the basis of confidently manipulable objects in the future. When things get sticky, or thinking breaks down, it is sensible to move down the spiral to reach some confidently manipulable examples from which to re-ascend. This is basically what Hilbert is reported to have used as his 'method' (Courant, 1981)

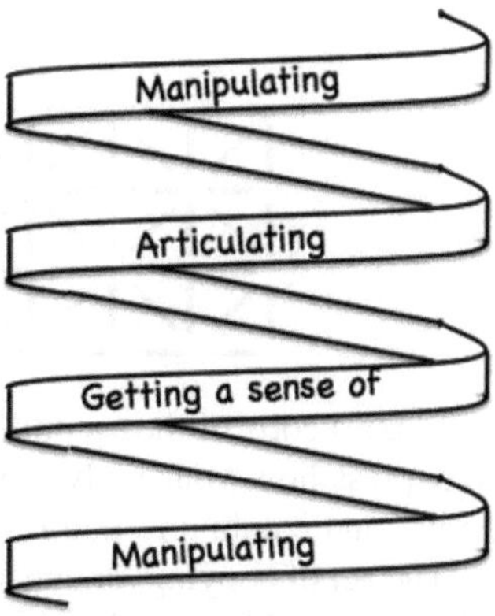

Since encounters with number, from the earliest moments, effectively draws on or makes use of the powers that enable abstraction and generality, working on getting learners to express generality in words, frequently, whenever appropriate, makes an important contribution to the developing of mathematical thinking. Indeed, you cannot appreciate and comprehend arithmetic without encountering the general (Hewitt, 1998).

6.3.2 *Routes into Symbols*

This section describes a collection of pedagogic strategies and didactic tactics which have been used to ease learners into the use of letters to denote the as-yet-unknown or the general. A plausible conjecture is that it is the sudden introduction of 'letters in place of numbers' which, for learners unused to denoting the as-yet-unknown or the as-yet-unspecified, triggers refusal to cooperate in algebra, or, for many who appear to cooperate, brings down the portcullis on pursuing mathematics because of the meaninglessness of symbol manipulation.

6.3.2.1 Watch What You Do and Say What You See

When seeking how to locate and/or extend a repeating geometrical pattern, or a numeric pattern with some growth structure, it is often useful to 'do an example', preferably a non-trivial example, or even to 'do' several examples. This has been the practice since recorded time! While drawing or calculating, it can be useful to pay attention to what your body wants to do (I use the slogan Watch What You Do or WWYD as a catch to remind me). For example, shown below are two configurations of squares made up of sticks, the first showing three rows of four columns and the second, four rows of six columns.

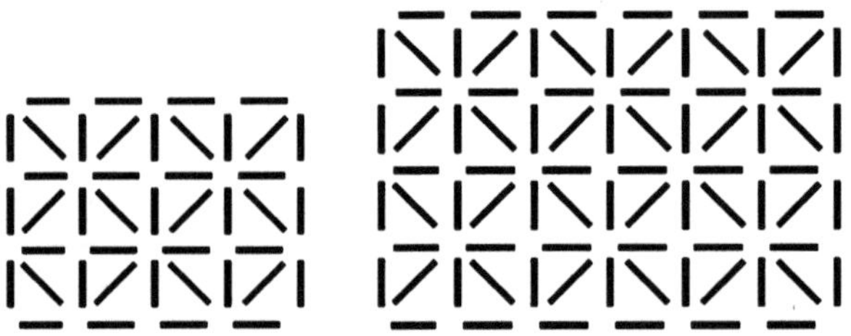

Make a copy of the second, watching how your body does the drawing. Then try to express how your body worked as a rule for how to draw a configuration with r rows and c columns, and how to count the number of sticks required.

The act of copying, or constructing your own instance, often leads to recognition of structure which can then be expressed verbally. Once refined, this provides a way to count the number of elements which can then be recorded using succinct symbols. For example, locating features in the first diagram which relate to three-ness and four-ness for which the same features in the second diagram relate to four-ness and six-ness is usually an acknowledgement by cognition of bodily awareness.

Note that the two 'examples' provided are not sequential, and do not start at 'the beginning'. It took me a long time to realise that always offering the first few terms of a sequence as examples was blocking learners' opportunities to use their own powers.

It is often the case that our bodies, our automatic functioning, locks into a pattern. For example, if invited to copy and extend the following for another nine rows,

$1 \times 7 = 7$	$7 \times 1 = 7$	$7 \div 1 = 7$
$2 \times 7 = 14$	$7 \times 2 = 14$	$14 \div 2 = 7$
$3 \times 7 = 21$	$7 \times 3 = 21$	$21 \div 3 = 7$

most children will quite spontaneously follow a flowing pattern downward, making use of the natural numbers and the invariants in each column. Anne Watson (2000) coined the expression 'going with and across the grain' to summarise what is made available to be learned in such a situation. To complete the mechanical part of the task, go with the grain, following the downward flow; to make sense of it, ask yourself what is changing and what is invariant, and how the three statements in a row relate to each other. This is 'going across the grain', revealing the structure, just as when you saw across the grain of a log you reveal the fibrous structure of the tree from which it came.

The slogan Say What You See (SWYS) can serve as a reminder to get learners to do articulate what they notice, first to a neighbour or group in which they are working, and then in plenary, where what is noticed can be recorded and organised. Once integrated into a learner's functioning, SWYS and WWYD can be powerful aides to detecting and expressing structure.

6.3.2.2 Tracking Arithmetic

Tracking Arithmetic is a label for the act of following one or more numbers through a sequence of calculations, in order to see what their role is, their influence, their contribution to the result. In other words, it leads directly to perceiving structural relationships and expressing generality. An especially powerful example is given by the following collection of tasks.

THOANs

Think of a Number 'games' have been played for hundreds, perhaps thousands of years. A simple version is the following:

Think of a (positive whole) number; add two; multiply by the number you first thought of; add one; take the (positive) square root (I can assure you that if you started with a positive whole number you will have a whole number square root). Subtract the number you first thought of. Your answer is 1.

Offered a sequence of these, perhaps using only addition and subtraction, children soon want to know how it is done, and to try it themselves. Tracking arithmetic reveals the underlying idea:

Start with 7. Add 2 to get not 9 but $7+2$. Multiply by the number you first thought of to get $7(7+2)$. Now add 1 to get $7(7+2)+1$. I can do the arithmetic to discover 64 whose square root is 8, but I want to see that 8 in terms of the 7, and I can see that $7(7+2)+1=7\times 7+2\times 7+1=(7+1)(7+1)$, so the square root is $7+1$. Subtracting the number first thought of yields 1 as claimed. The 7 has been made to disappear! Now replace every instance of the starting 7 with a cloud (it might be that 7 also shows up spontaneously in the calculation so one has to be wary):

Using a cloud, which draws upon learners' experience of cartoons, has in my experience enabled algebra–refusers in secondary school both to engage and to act algebraically, blissfully unaware that they have been 'doing algebra'. A good deal of the energy exhibited by learners who have chosen to become algebra–refusers lies in their not knowing what the letters of algebra refer to. As Mary Boole put it, the use of algebra is to *free people from bondage* (Tahta, 1972, p. 55; italics in original), by which she means bondage by and to the particular.

A particularly effective use of tracking arithmetic can be made by tracking all numbers in the following task.

Grid Sums

Write down four numbers in a two-by-two grid (as in the example)	5 3 7 4
Record the products along the rows and the products down the columns Now add the column sums and subtract both the row sums	5 3 15 7 4 28 35 12
The result in this case is $35+12-15-28=4$ Now choose numbers for a new grid so as to make the result equal to 3 (or any other pre-assigned number!)	

Most people start trying numbers and doing calculations. Tracking arithmetic reveals an underlying structure:

The row sums are 5×3 and 7×4; the column sums are 5×7 and 3×4, so the result is

$$\begin{aligned} 5 \times 7 + 3 \times 4 - 5 \times 3 - 7 \times 4 &= (5 \times 7 - 5 \times 3) + (3 \times 4 - 7 \times 4) \\ &= 5 \times (7-3) + (3-7) \times 4 \\ &= 5 \times (7-3) - (7-3) \times 4 \\ &= 5 \times (7-3) - 4 \times (7-3) = (5-4) \times (7-3) \end{aligned}$$

The result is the product of the differences along the diagonals! Once that structure is recognised, it is easy to achieve any pre-assigned result, whereas without it, achieving a specified number can be really challenging. Of course if you are already familiar and confident with using letters, you can do it 'algebraically', but Tracking Arithmetic is available even if you do not yet have algebraic facility. Notice however that you do need some general arithmetic facility, which is why it is worth, early on in arithmetic, drawing attention to the properties of arithmetic such as commutativity, associativity and distributivity.

As an extension, why does the result stay the same if I choose two additional numbers, add the first number to the upper left and lower right cells, and subtract the second number from the lower left and upper right numbers?

Since no task is an island complete unto itself (Mason, 2010), how might this task be altered or extended? It turns out that it is not obvious how to extend the idea to a three-by-three grid. However, there is a variation which might be somewhat surprising.

Reading clockwise from the upper left corner, form two two-digit numbers. In my case I get 53 and 47. Do the same counterclockwise to get 57 and 43. Now form the difference of the products: $53 \times 47 - 57 \times 43 = 40$

5	3
7	4

Adjusting the grid by subtracting say 1 from the main diagonal numbers and adding say 2 to the off diagonal numbers gives the grid shown, and $45 \times 39 - 49 \times 35 = 40$ as well. Could this be a coincidence?

4	5
9	3

Tracking arithmetic on the original grid shows that

$$\begin{aligned}53 \times 47 - 57 \times 3 &= (50 + 3) \times (40 + 7) - (50 + 7) \times (40 + 3)\\ &= (50 \times 40 + 50 \times 7 + 3 \times 40 + 3 \times 7)\\ &\quad - (50 \times 40 + 50 \times 3 + 7 \times 40 + 7 \times 3)\\ &= (50 \times 7 + 3 \times 40) - (50 \times 3 + 7 \times 40)\\ &= 50 \times (7 - 3) + (3 - 7) \times 40\\ &= (50 - 40) \times (7 - 3)\\ &= 10 \times (5 - 4) \times (7 - 3).\end{aligned}$$

It is immediately evident then that adding or subtracting the same thing to/from the main diagonal numbers makes no difference, nor does adding or subtracting the same number to/from the off diagonal elements. Furthermore, the result must always be ten times the result of the previous calculation using the grid numbers. To 'see' this for oneself requires only locating the 5, 4, 7 and 3 in the grid itself, and realising (making real for oneself) that the digits are acting as placeholders and can be changed.

Tracking arithmetic provides an intermediate stage between using arithmetic with particular numbers and using letters for as-yet-unspecified numbers (our ignorance). As such it is a *didactic* tactic (Mason, 2002b): it is particularly useful and applicable to generating experience of algebraic thinking. I know of several tasks which enable students to work with generality without having to call upon the particular at all (see Sect. 6.4 for another example) and there must be many more.

6.3.2.3 Acknowledging Ignorance

Mary Boole (see Tahta, 1972, p. 55) suggested that algebra arises from 'acknowledging ignorance'. When you recognise that you do not know 'an answer' you can acknowledge that fact by using a symbol (a little cloud is particularly effective) to denote what is not (yet) known. You can then use that cloud to express what you do know about it, and this will usually lead you to some constraints on the generality of 'cloud' in the form of equations or inequalities. This is what Isaac Newton (1683) thought was so elementary! Of course there are circumstances where this does not help, but these are rare in school algebra examinations!

I have written down two numbers whose sum is one. I square the larger and add the smaller; I square the smaller and add the larger. Which of my two numbers will be the larger?

Notice that strong force to try a particular example. Choosing 0 and 1, or 1/2 and 1/2 is not very revealing. The fact that the two calculations always give the same result is, at least at first, a little surprising. Acknowledging our ignorance and denoting one of the numbers by  and the other by 1- is already using the cloud to express what you know, namely that they sum to 1. Now the calculations can be done using the cloud. If learners are not yet ready for manipulating cloud, than tracking arithmetic can be used:

Try 7 as one number, and $1-7$ as the other (notice that any calculation involving 7 is indicated but not carried out). Then the two calculations give
$7^2 + (1-7) = 7^2 - 7 + 1$
and $(1-7)^2 + 7 = 1^2 - 2 \times 1 \times 7 + 7^2 + 7 = 7^2 - 7 + 1$
So the two calculations are equal in this instance.

Treating the 7 now as a place holder rather than as a particular number, perhaps at first replacing it by a little cloud, confirms that the two calculations always give the same result. It is worth pausing and contemplating the scope or range of generality. The 7, or the cloud, can be replaced by *any* number you can think of, or indeed numbers you cannot even think of or which have never previously been thought of!

A useful task for emphasising the scope and range of generality involves variants of the following:

Write down a number between 3 and 4.
Now write down a number between 3 and 4 but which no one else in the room will write down.
Now write down a number between 3 and 4 but which no human being is ever likely to have written down.

The second version draws attention to the range of possible choices. The third version sharpens awareness that there are more numbers than human beings have ever used! The idea is to draw attention to the range of possible variation, the scope of generality.

Note that in a task like this there is an opportunity to get a learner to choose what the difference will be. That way they have a sense of both the 3 and the 4 as place holders for a dimension of variation (a generality) as well as experiencing greater commitment to the task because they have participated in making a significant choice.

A related tactic is to make a guess, and then check whether your guess is correct. If you can check the correctness of a guess, then you can use tracking arithmetic to follow the guess through the checking process, using a little cloud or other token, and end up with equations and inequalities which express the constraints on the generality of your 'guess'. The method of *false position* which pervades arithmetic books up until the nineteenth century is based on a way of making use of one or more guesses and the errors they give rise to when checking them, purely arithmetically, in order to determine the correct answer. This only works when the calculation is linear (one trial guess) or quadratic (two trial guesses), and rarely did authors of textbooks give any criteria for knowing whether one guess or two were required!

6.3.2.4 Word Problems

It has already been noted that if you can check the answer to a question, you can usually set it out algebraically, by tracking arithmetic: following your proposed answer through the calculations without losing track of it. Then you can set up the constraints on it as equations or inequalities, and perhaps even solve them to find the correct answer. This applies particularly to 'word problems'. But asking learners to 'solve' word problems is likely to be met with hostility, whether cognitive, affective or enactive, and perhaps all three. By contrast, the notion of 'burying the bone' (Watson & Mason, 2005), of getting students to try to construct a problem that they can do themselves but that will challenge colleagues, perhaps even the teacher, can be used to increase engagement and disposition. This actually mirrors the competitions in Italy in the sixteenth century involving Nicolo Tartaglia and Girolamo Cardano (MacTutor Website) which brought to light the formula for solving a cubic equation! Invoking the theme of 'doing and undoing', by asking learners to construct problems 'like these' which will challenge others, puts learners in the role of constructors, or meaningful agents. They may even come to appreciate the complexity of setting problems which will enable others to display their understanding, such as examiners. The more that learners get to make significant mathematical choices, the more likely they are to appreciate the tasks they are set, because they know how they are constructed and for what purpose.

Word problems can also be used to challenge people to find a solution without using algebra! Algebra becomes a backstop, a place of last resort. Meanwhile they are exercising their mathematical thinking in trying to find a purely arithmetic resolution. Then they can use Tracking Arithmetic to express a general formula for all problems of 'that type'. This is how Newton (1683) presented his solutions: he solved a particular, then the general, and then showed that the particular was an instance of the general.

6.4 Reasoning Without Numbers

It is well worth while looking out for opportunities for learners to reason without having to work with numbers, especially if some or all have already developed a reluctance to master arithmetic.

6.4.1 Magic Square Reasoning

Imagine that the initial three-by-three square is covering up some three-by-three magic square. It doesn't matter which one. The fact that it is a magic square means that the sum of the numbers in any row, any column or either diagonal is the same. So in particular, the sum of the numbers in the cells in the first row is the same as the sum of the numbers in the first column.

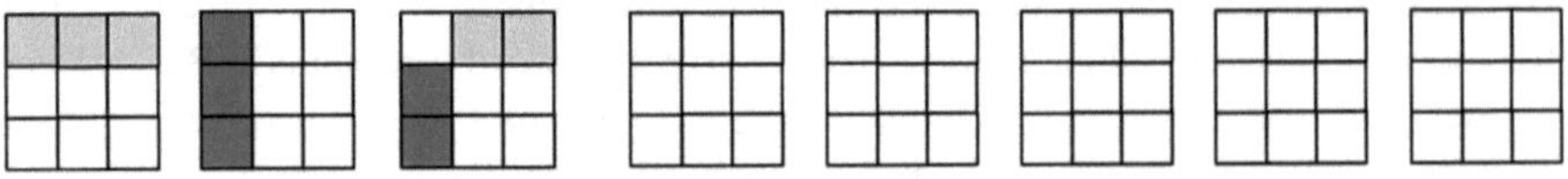

The sum of the light-shaded cells in the first grid is the same as the sum of the dark-shaded cells in the second grid, and because these would overlap, as shown in the third grid the sum of the dark-shaded cells must be the same as the sum of the light-shaded cells in the third grid.

On the remaining grids, shade in sets of cells so that the sum of the dark-shaded cells *must be* the same as the sum of the light-shaded cells.

In the following grids, show why the sum of the dark-shaded cells must be the same as the sum of the light-shaded cells.

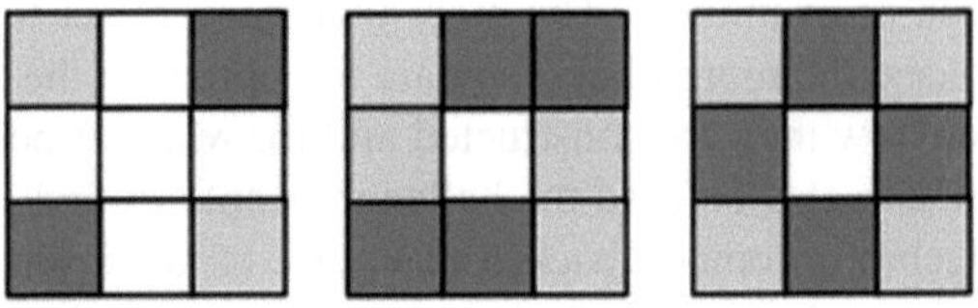

Notice that you do not need to know any of the numbers ... the reasoning is all about rows, columns and diagonals with overlaps removed. However, it is not always easy to see how to achieve someone else's configuration. Things become even more challenging and hence interesting when you move to four-by-four or larger magic squares.

The power of the reasoning using overlaps is that the results apply to any magic square whatsoever, and yet numbers are not actually used. Learners find themselves thinking structurally, algebraically. Care is needed however, that learners keep in mind that the patterns they are using involve rows, columns and diagonals only, and a balance between the number of these in one colour and the number in the other colour, because these all have the same sum. In an experiment with children aged 11 it turned out that making patterns of colours dominated attention, and they lost the idea of using only rows, columns and diagonals and eliminating overlaps (Mason, Oliveira, & Boavida, 2012).

6.5 Reasoning About Numbers

Getting learners to reason about numbers, rather than doing arithmetic with them can encourage arithmetic-refusers to engage even though numbers are involved. For example,

I am about to subtract the number represented by the cloud (it is a number that someone is thinking about) from the number represented by the box (it is also a number that someone else is thinking about).

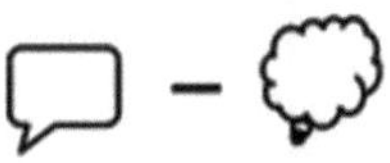

However, just before I do the subtraction, someone comes along and adds 1 to both of the numbers. How will the subtraction result change?

The invariance is both intuitive and readily justified. How can this task now be extended and developed? What aspects could be varied? Variation theory (see Marton, 2015; Marton & Booth, 1997) suggests that what is available to be learned is what has been varied in recent time and space. Teaching is seen as fundamentally about opening up dimensions of possible variation so that learners not only become aware of possibilities, but integrate into their functioning the action of considering what can be varied, and over what range and with what constraints ('range of permissible change': see Watson & Mason, 2005).

In this task, the adjustment by 1 is a dimension of possible variation, leading to the recognition that the same adjustment to both numbers will make no difference. Opening up the constraint that the adjustments must be the same leads to further insight. Note the parallel with the grid-sums task in Sect. 6.3.2.2). Altering subtraction to addition, to division or multiplication reveals similarities and differences in the language and the actions that preserve an invariance.

6.6 Generally Looking Beyond the Particular

Extending and varying, informed by variation theory are just as vital as getting answers to some task. The learner who arrives at a test or examination and who has treated every task as isolated has to deal with each test item in its particularity, whereas the learner who has extended and varied, who has developed a rich space of examples and of ways to augment and modify examples, is likely to recognise the type of task and to have possible actions become available almost automatically. I have long encouraged learners about to take an exam to set their own exam and send it to the examiner, engaging in dialogue about what is reasonable and what is challenging, and why. In that way learners become acquainted with what testing is about, and develop their facility by extending and varying for themselves.

For example, the task One Sum presented earlier can be extended and varied in several ways, but most easily when the situation is depicted.

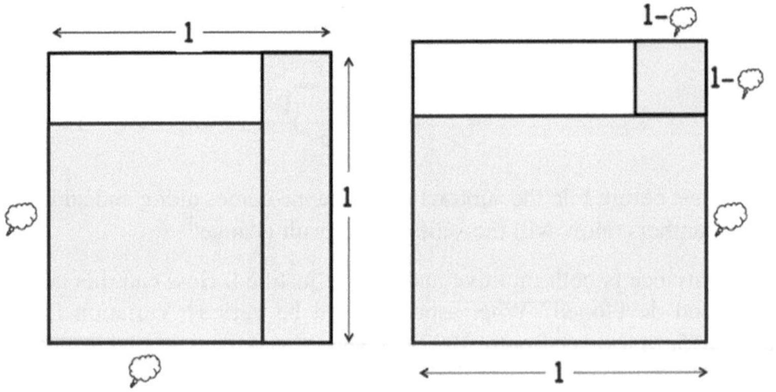

As often happens in mathematics, finding two or more ways to express the same thing can be enlightening and productive. Here the shaded area can be broken down in two ways, and this leads to other possibilities, taking the number of numbers adding to one as a dimension of possible variation, and taking the two-dimensionality as a dimension of possible variation.

Use the two diagrams below to express generalisations of the one-sum relationship.

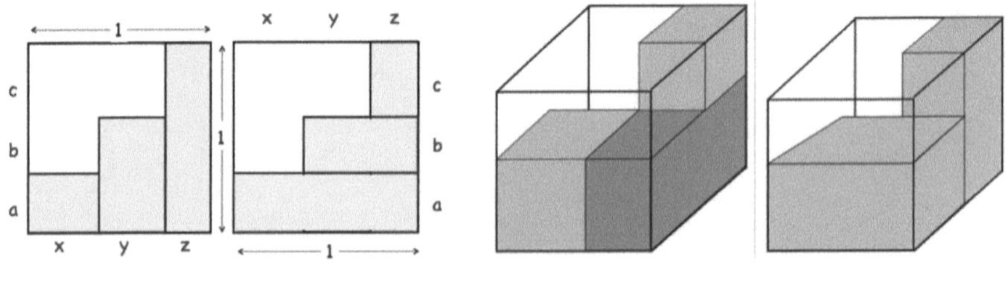

Working on expressing these involves both algebraic thinking, and shifting of attention back and forth from recognising relationships in the particular diagrams, and perceiving these as instantiations of properties (Mason, 2001).

6.7 A Word of Caution

Just because some pattern or relationship can be extended, it does not mean that it is true. Put another way, every expression of generality starts life as a conjecture. It must be tested and justified. Even with elementary repeating patterns, care must be taken not to give learners the mistaken impression that whatever they think might be true, will be true.

6.7.1 Repeating Patterns

The following pattern is made from repeating a block of letters. Extend the sequence for yourself so that the repeating block continues to repeat.

AAABAA

Of course there are several ways: the repeating block can be any of *AAAB*, *AAABA*, *AAABAA*, assuming that the generating pattern appears at least once. To make the pattern unique, it is mathematically necessary to know that the repeating pattern generating the sequence appears at least twice (Mason, 2014). For example,

AAABAAAABAAAA

with the claim that there are at least two copies of the repeating pattern, is uniquely identifiable and therefore extendable.

6.7.2 Power Sums

It is well known that $3^2 + 4^2 = 5^2$, but not so well known that $3^3 + 4^3 + 5^3 = 6^3$. Having checked this, it is hard to resist trying extensions ... but they don't work!

The 'obvious' or 'natural' generalisation turns out to be false. The point is that the first two facts are not presented in a structural form which actually extends. If there is a suitable extension, some structural underpinning is required. That is why whenever learners are asked to extend a sequence, or to count the number of objects needed to make a term in a sequence, they must first be asked to articulate what the structural underpinning is that generates the sequence.

6.7.3 Structural Foundations

Consider, for example, the first two terms of a picture sequence:

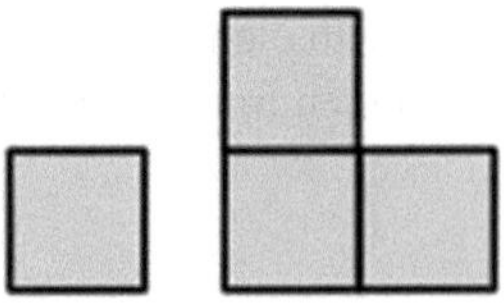

The third term could be any of the following

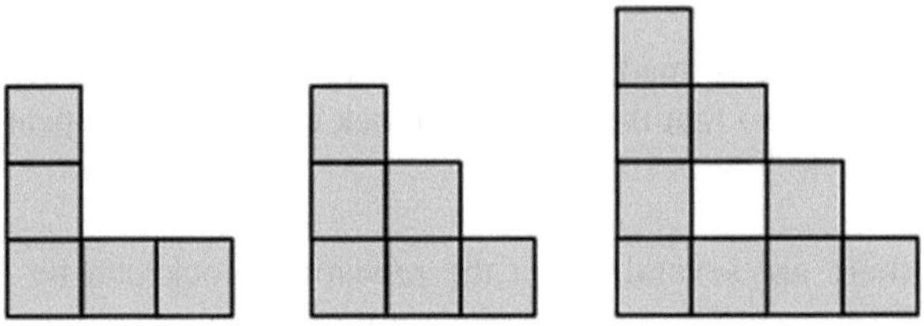

not to say something completely different. Without specifying how the diagrams are to be constructed, it is not possible to count the squares needed to make the *n*th picture.

Use of pedagogic strategies such as getting learners to consider, having resolved one problem, to consider the range of tasks they can solve similarly, and getting them to change what is given and what is sought (a manifestation of the mathematical theme of *doing & undoing*) not only engages learners more deeply, but also offers them some actions to make use of for and by themselves, when studying, and when interacting with the world generally. Thus in the study by Jo Boaler (1997) learners at Phoenix Park, where mathematical thinking was encouraged through work on extended tasks, recognised the role of mathematics outside of the classroom in ways that students taught more traditionally as a sequence of procedures to be mastered did not.

Not only does extending or varying aspects of a task, exploring possible dimensions of variation, increase engagement with tasks, and not only does it provide ways for quicker learners to remain engaged, it is the very heart of mathematics, building up rich example spaces on which learners can draw in the future. One important way to augment the affectivity of wanting to engage is to take every opportunity to get students to make significant mathematical choices for themselves: what examples they work on, what letters they use to stand in for an as-yet-unknown or a yet-to be decided unknown, whether to specialise or to work with the general, and so on.

Even when you cannot see how to extend or vary, it is worthwhile trying. For example, I came across the following task in Pólya (1954, Ex. 7, pp. 117–118) and included it in Mason, Burton, and Stacey (1982, p. 169). Pólya noticed it in our book and asked why we had associated his name with it, which was because we got it from him!

6.7.4 Pólya Strikes Out

Write out the natural numbers in a sequence	1 2 3 4 5 6 7 8 9 10 11 12 13 14 ...
Circle every other number	1 ⓶ 3 ⓸ 5 ⓺ 7 ⓼ 9 ⑩ 11 ⑫ 13 ⑭ ...
Form the cumulative sums of the uncircled numbers	1 4 9 16 25 36 ...

Not too surprisingly, we get the square numbers. If instead you begin by circling every third number, forming cumulative sums, then circle every second number in this, and form the cumulative sums, you get another recognisable sequence. Repeating this sort of action continues to reveal recognisable sequences. Try as I might I could not get beyond a simple generalisation. Then John Conway and Richard Guy (1996, pp. 63–65) found it in a paper of Moessner (1952: see Conway and Guy 1996, p. 89) and generalised it extensively. They noticed that if instead of using 'every-something' as the circling rule, you circle each number in a triangular-number position, repeatedly, then the first circled numbers in each row form another familiar sequence, and that is just the beginning!

The slogans 'be wise, generalise' (attributed to Piccayne Sentinel: see MAphorisms) and 'there is always something more to discover in the way of connections and relationships' are part of a mathematician's creed, though it must also be noted that Paul Halmos (1975) decried the effect on graduate students of using the first without also being aware of instantiations of those generalisations, and of where in mathematics they might be relevant. William Blake also decried generalisation, claiming that 'to generalize is to be an idiot'. I take the more balanced view that generalisation and instantiation in the particular are both important, in fact are inescapably intertwined, and that to focus on one without the other is indeed to be an idiot.

6.8 Classroom Ethos

For mathematical thinking to take place effectively, there has to be a *conjecturing atmosphere* (Mason, Burton, & Stacey, 1982/2010, pp. 64, 233). This is so much a part of mathematicians' practice that books often do not bother mentioning it. Yet it is fundamental. In such a classroom ethos, those who are confident about a question or a task listen to what others have to say, while those who are not confident try to say what they can. Things are said (by learners, by the teacher) in order to get them outside of the 'tumble-dryer' mind in which ideas get mixed up, change, and develop, even in mid expression. Things are said as conjectures in order to consider them dispassionately. Then, as George Pólya (1965) put it, 'you must not believe your conjecture'.

Instead of disagreeing with what someone says, or telling them they are wrong, in a conjecturing atmosphere you might ask about how what was said plays out in . . . (and here an example, perhaps a potential counter-example is offered). Learners quickly find that asking someone to repeat what they said is less productive than trying to say what you think you heard, and asking for validation and clarification.

6.8.1 *Increasing Sums*

Consider the portion of Pascal's triangle shown below, and convince yourself you know how to extend it to the right and down.

1	1	1	1	1	1	1	1	1	1	1	1
1	2	3	4	5	6	7	8	9	10	11	12
1	3	6	10	15	21	28	36	45	55	66	78
1	4	10	20	35	56	84	120	165	220	286	364

Now group as shown below

1 = 1	1 + 1 = 1 + 1	1 + 1 + 1 = 1 + 1 + 1			
1 + 2 = 3	4 + 5 + 6 = 7 + 8	9	10	11	12
1 + 3 + 6 = 10	15 + 21 + 28 + 36 = 45 + 55	66	78		
1 + 4 + 10 + 20 = 35	56 + 84 + 120 + 165 + 220 ≠ 286 + 364				

Say What You See in this diagram. Take your time. It might even help to make a copy for yourself, and Watch What You Do. Ask yourself what is invariant (not just objects, but relationships), and what is changing and in what way(s). What is the same and what is different about each row, about the groupings in each row, about the groupings in a sequence of rows?

The first row groupings seem trivial, but in retrospect from the second and third rows they make sense. But the fourth row displays a counter-example to a common conjecture! A generalisation, an expression of generality, is always a conjecture until it can be justified! Do the first groupings in each row continue? Why then don't the second groupings in each row continue?

Note the pedagogic strategies instantiated in the follow-up part of the task.

6.9 Summary

Drawing on more than 50 years of working with others to develop mathematical thinking, it seems clear to me that there is no royal road to teaching, no single track to pedagogy, no magic sequence of tasks that will achieve the transformation in thinking algebraically sought after for so many centuries by so many teachers. Quite the contrary, it is all about sensitivity to individuals and to groups of individuals. It is all about teaching as a caring profession: caring for learners and caring for the subject matter, which requires maintaining a balance between the two and not going to extremes. As an old adage has it 'every stick has two ends'. It is all about responding to particular situations with access to a rich repertoire of pedagogic strategies and didactic tactics. It is about nurturing like a gardener rather than managing an assembly line.

Developing facility in manipulating algebra is actually straightforward once confidence and interest in working with generalities has been captured.

In this chapter I have offered some pedagogic strategies, some didactic tactics, and some tasks through which to encounter these, which, if handled sensitively and carefully, not as one-off events but as a classroom ethos, a way of working with others, could make a difference to succeeding generations.

References

Bednarz, N., Kieran, C., & Lee, L. (Eds.). (1996). *Approaches to Algebra: Perspectives for research and teaching*. Dordrecht: Kluwer.

Boaler, J. (1997). *Experiencing school mathematics: Teaching styles, sex and setting*. Buckingham: Open University Press.

Bruner, J. (1966). *Towards a theory of instruction*. Cambridge: Harvard University Press.

Cai, J., & Knuth, E. (2011). *Early algebraization: A global dialogue from multiple perspectives*. Heidelberg: Springer.

Chevallard, Y. (1985). *La Transposition Didactique*. Grenoble: La Pensée Sauvage.

Chick, H., Stacey, K., Vincent, J., & Vincent, J. (Eds.). (2001). *The future of the teaching and learning of algebra*. Proceedings of the 12th ICMI Study Conference, University of Melbourne, Melbourne.
Conway, J., & Guy, R. (1996). *The book of numbers*. New York: Copernicus.
Courant, R. (1981). Reminiscences from Hilbert's Gottingen. *Mathematical Intelligencer, 3*(4), 154–164.
Davis, B. (1996). *Teaching mathematics: Towards a sound alternative*. New York: Ablex.
Davydov, V. (1990). *Types of generalisation in instruction* (Soviet studies in mathematics education, Vol. 2). Reston: NCTM.
Dougherty, B. (2008). *Algebra in the early grades*. Mahwah: Lawrence Erlbaum.
Gattegno, C. (1988). *The mind teaches the brain* (2nd ed.). New York: Educational Solutions.
Giménez, J., Lins, R., & Gómez, B. (Eds.). (1996). *Arithmetics and algebra education: Searching for the future*. Barcelona: Universitat Rovira i Virgili.
Halmos, P. (1975). The problem of learning to teach. *American Mathematical Monthly, 82*(5), 466–476.
Hewitt, D. (1998). Approaching arithmetic algebraically. *Mathematics Teaching, 163*, 19–29.
Kaput, J., Carraher, D., & Blanton, M. (2008). *Algebra in the early grades*. Mahwah: Lawrence Erlbaum.
MacTutor Website. Retrieved from http://www-history.mcs.st-and.ac.uk/index.html
MAphorisms. Retrieved October, 2015, from www.math.ku.dk/~olsson/links/maforisms.html
Marton, F. (2015). *Necessary conditions for learning*. Abingdon: Routledge.
Marton, F., & Booth, S. (1997). *Learning and awareness*. Hillsdale, MI: Lawrence Erlbaum.
Mason, J. (2001). Teaching for flexibility in mathematics: Being aware of the structures of attention and intention. *Questiones Mathematicae, 24*(Suppl 1), 1–15.
Mason, J. (2002a). *Researching your own practice: The discipline of noticing*. London: Routledge-Falmer.
Mason, J. (2002b). *Mathematics teaching practice: A guidebook for university and college lecturers*. Chichester: Horwood.
Mason, J. (2010). Attention and intention in learning about teaching through teaching. In R. Leikin & R. Zazkis (Eds.), *Learning through teaching mathematics: Development of teachers' knowledge and expertise in practice* (pp. 23–47). New York: Springer.
Mason, J. (2014). Uniqueness of patterns generated by repetition. *Mathematical Gazette, 98*(541), 1–7.
Mason, J., Burton, L., & Stacey, K. (1982). *Thinking mathematically*. London: Addison Wesley.
Mason, J., Graham, A., Pimm, D., & Gowar, N. (1985). *Routes to, roots of Algebra*. Milton Keynes: The Open University.
Mason, J., & Johnston-Wilder, S. (2004). *Designing and using mathematical tasks*. Milton Keynes: Open University.
Mason, J., Oliveira, H., & Boavida, A. M. (2012). Reasoning reasonably in mathematics. *Quadrante, XXI*(2), 165–195.
Mason, J., & Pimm, D. (1984). Generic examples: Seeing the general in the particular. *Educational Studies in Mathematics, 15*(3), 277–290.
Mason, J., & Sutherland, R. (2002). *Key aspects of teaching algebra in schools*. London: QCA.
Moessner, A. (1952). Ein Bemerkung über die Potenzen der natürlichen Zahlen. *S.–B. Math.-Nat. Kl. Bayer. Akad. Wiss., 29* (MR 14 p353b).
Newton, I. (1683) in D. Whiteside (Ed.). (1964). *The mathematical papers of Isaac Newton* (Vol. V). Cambridge: Cambridge University Press.
Nunes, T., & Bryant, P. (1996). *Children doing mathematics*. Oxford: Blackwell.
Nunes, T., Bryant, P., & Watson, A. (2008). *Key understandings in mathematics learning*. Retrieved October, 2015, from www.nuffieldfoundation.org/key-understandings-mathematics-learning
Open University. (1982). *EM235: Developing mathematical thinking*. A distance learning course. Milton Keynes: Open University.

Pólya, G. (1954). *Mathematics and plausible reasoning* (Induction and analogy in mathematics, Vol. 1). Princeton: Princeton University Press.

Pólya, G. (1965). *Let us teach guessing* (film). Washington: Mathematical Association of America.

Schmittau, J. (2004). Vygotskian theory and mathematics education: Resolving the conceptual-procedural dichotomy. *European Journal of Psychology of Education, 19*(1), 19–43.

Tahta, D. (1972). *A Boolean anthology: Selected writings of Mary Boole on mathematics education*. Derby: Association of Teachers of Mathematics.

Ward, J. (1706). *The young mathematicians guide, being a plain and easy Introduction to the Mathematicks in Five Parts*. Thomas Horne: London.

Watson, A. (2000). Going across the grain: Mathematical generalisation in a group of low attainers. *Nordisk Matematikk Didaktikk (Nordic Studies in Mathematics Education), 8*(1), 7–22.

Watson, A., & Mason, J. (2005). *Mathematics as a constructive activity: Learners generating examples*. Mahwah: Erlbaum.

Chapter 7
Algebra as Part of an Integrated High School Curriculum

James T. Fey and David A. Smith

Abstract Traditional high school mathematics curricula in the United States devote 2 years almost exclusively to development of student proficiency in the symbolic manipulations required for solving algebraic equations and generating equivalent algebraic expressions. However, recent design experiments have shown that a focus on functions, mathematical modeling, and computer algebra tools enables effective integration of algebra with the other core strands of high school mathematics.

Keywords Integrated curriculum • School algebra • Functions • Problem-based learning • Mathematical modeling • Applications

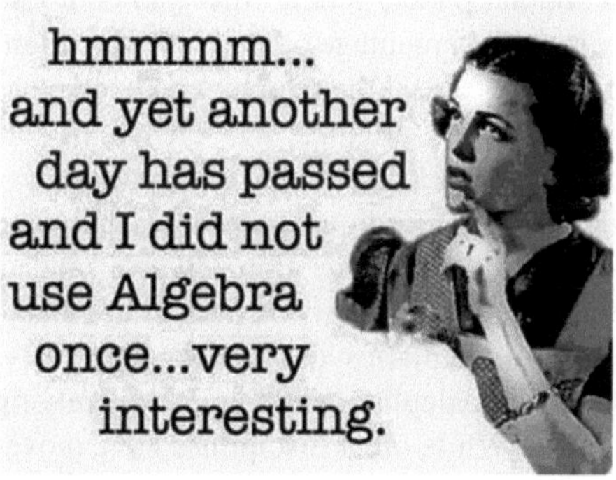

J.T. Fey (✉)
University of Maryland, College Park, MD, USA

13804 Beacon Hollow Lane, Wheaton, MD 20906, USA
e-mail: jimfey@umd.edu

D.A. Smith
Duke University, Durham, NC, USA

S. Stewart (ed.), *And the Rest is Just Algebra*, DOI 10.1007/978-3-319-45053-7_7

The figure at the left is a meme circulating on the Internet that summarizes nicely the public perception of the importance of algebra as it is taught in most American schools. The statement may be true of every other school subject save English, but it gets a chuckle only for algebra.

Long-standing tradition in American education calls for organization of the high school mathematics core curriculum in a layer cake of 3 year-long single-subject courses—elementary algebra, geometry, and advanced algebra. The algebra courses that dominate this curriculum emphasize training students in what Robert Davis once characterized as a 'dance of symbols'—a collection of procedures for manipulating symbolic expressions, equations, and inequalities. The applications of those symbol manipulation rules are commonly limited to an array of classic word problems of dubious authenticity, and precious little attention is given to topics with easily demonstrated practical importance such as probability, statistics, and modern discrete mathematics, much less the process of mathematical modeling that is central to contemporary applied mathematics.

This dominant structure of American high school mathematics curricula appears to have evolved during the late nineteenth and early twentieth century as mathematics courses that had been in the curricula of prominent colleges were transformed into admission prerequisites for those institutions (Jones & Coxford, 1970; Kilpatrick & Izsak, 2008). We, the authors, were high school students in the 1950s, when it could be argued that "classical" algebra and geometry were the subjects that could be taught in secondary schools because they could be mastered with the tools then available: paper, pencil, ruler and compass (for geometry), and brain power. But only a small percentage of students were highly successful in these courses, and they tended to be "us," the people who became college and university faculty members in mathematics, engineering, science, or education.

Half a century later, the world is a very different place. The problems to be solved are more challenging, our brain-extender tools are much more sophisticated, our school and college populations are more diverse, and our knowledge-based economy is no longer dependent primarily on agriculture and manufacturing. But in most places our school mathematics curriculum has evolved only marginally, "allowing" use of electronic calculators and no longer relying solely on Euclid as the definition of geometry. While other disciplines have moved on,[1] the "standard" secondary mathematics curriculum has much in common with the Saber-Tooth Curriculum (Benjamin, 1939/2004).

Prompted by striking findings from a series of recent international studies of mathematics teaching and learning, American mathematics educators have explored different ways of thinking about the high school curriculum. In particular, efforts such as the Interactive Mathematics Project (Fendel, Resek, Alper, & Fraser, 2015) and the Core-Plus Mathematics Project (Hirsch, Fey, Schoen, Hart, & Watkins, 2014) have developed and tested American versions of the most common

[1] In the 1950s, the importance of DNA was not well known in biology, black holes were not known to exist, and tectonic plates were still considered heresy by earth scientists.

international model that advances student understanding of all major content strands in each year of high school study. These so-called integrated or standards-based[2] mathematics curricula give significant attention to a broader range of mathematical topics than traditional algebra-centric curricula, and they also pay explicit attention to developing student understanding and skill in mathematical processes such as problem solving, communication, reasoning, and connection of ideas.[3]

Proposals to broaden and integrate topics in high school mathematics challenge the centrality of traditional algebraic content. The focus of school algebra on formal procedures for manipulating expressions, equations, and inequalities is also challenged by emergence of technological tools such as graphing calculators and computer algebra systems. If every algebraic symbol manipulation can be performed quickly and accurately by universally available computer software, is it still important for students to spend 2 full years of high school study in pursuit of what is inevitably incomplete and fragile mastery of those same routines? What is the right agenda of algebra learning goals for students today?

7.1 New Perspective: Function as Fundamental Concept

One curricular approach that has shown great promise in several innovative integrated curriculum projects replaces the traditional focus on formal manipulation of symbolic expressions and equations with an emphasis on a different fundamental mathematical idea—*functions*. To see what a focus on functions might look like and how it can lead to productive development of still essential algebraic understandings and skills, consider a problem that requires mathematical modeling and reasoning:

A new professional sports league has a business problem:

What average ticket price will maximize operating profit of the league all-star game?

The situation involves several key variables—number of tickets sold, income from ticket sales, income from concession sales, operating costs, and operating profits—most of which depend ultimately on average ticket price. Market research and other business analyses could lead to function models for those dependencies. For example, we might come up with functions such as these:

[2] The term *standards-based* generally refers to curricula that embody recommendations of the 1989 National Council of Teachers of Mathematics *Curriculum and Evaluation Standards for Teaching Mathematics* and the 2000 *Principles and Standards for School Mathematics*.

[3] The notion that proficiency in mathematics includes certain *habits of mind*, as well as knowledge of specific facts, concepts, and procedural skills, has been reflected in all professional curriculum guidelines over the past quarter-century, most recently in the *Common Core State Standards for Mathematics*.

$$\begin{aligned} \text{Demand} : n(x) &= 5000 \;-\; 65x \\ \text{Income} : I(x) &= 5000x \;-\; 65x^2 \\ \text{Expenses} : E(n) &= 4n + 25,000 \\ E(x) &= 45,000 \;-\; 260x \\ \text{Profit} : P(x) &= -65x^2 + 5260x \;-\; 45,000 \end{aligned}$$

Algebraic notation and symbol manipulation are very useful in expressing such problem conditions, in calculations for finding derivatives, and in solving equations to find specific numeric values such as optimum ticket and break-even prices. In fact, for a conventional treatment of this problem, algebraic skills are essential.

But using the numeric, graphic, and symbolic tools provided by calculators and computers, one has access to very effective approximation strategies for solving equations and inequalities and even finding maximum or minimum values for functions. Furthermore, contemporary computer algebra systems will actually perform all required exact calculations.

$$\frac{d}{dx}\left(-65x^2 + 5260x - 45,000, x\right) \quad 5260 - 130x$$

$$\text{solve}(5260 - 130x = 0, x) \quad x = \frac{526}{13} \quad \text{or } x = 40.46$$

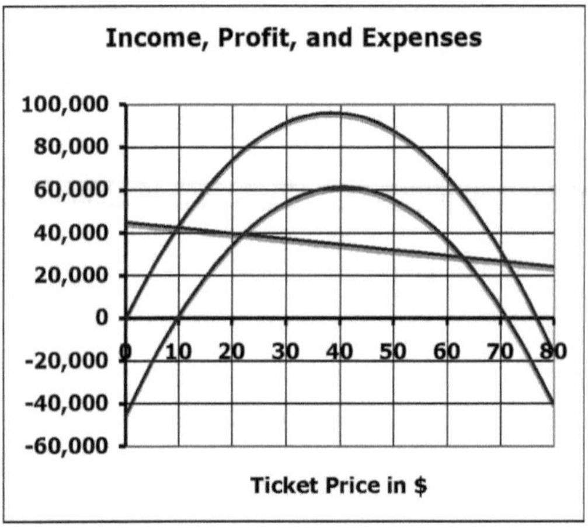

Price	Income	Expense	Profit
0	0	45,000	–45,000
10	43,500	42,400	1100
20	74,000	39,800	34,200
30	91,500	37,200	54,300
40	96,000	34,600	61,400
50	87,500	32,000	55,500
60	66,000	29,400	36,600
70	31,500	26,800	4700
80	–16,000	24,200	–40,200

The business analysis problem posed by planning for a sports league all-star game is typical of tasks encountered by students in calculus for the management sciences—a course for which high school algebra is assumed to be essential preparation. Success in the conventional version of that course certainly does require proficiency in writing and manipulating algebraic expressions and in solving equations. But the central concepts of calculus are functions and rates of change, and applying those concepts to realistic problems requires thinking about *variables*, *expressions*, and *equations* in different ways than the traditional approaches to elementary algebra emphasize. In applications of calculus, variables represent quantities that change over time or in response to change of other related variables. Equations show how variables are related. Expressions show how to calculate values of dependent variables. So instead of thinking about algebra as only a collection of symbol manipulation techniques for discovering fixed but unknown values of x, it makes sense to think of algebra as a way of expressing and reasoning about relationships between changing quantities. Techniques for solving equations and inequalities are helpful in finding answers to specific questions situated in the context of broader quantitative relationships.

Using functions—rather than symbolic expressions, equations, and manipulation—as the central organizing concepts for high school mathematics has a number of important payoffs.

- As the preceding example shows, viewing algebraic expressions and equations as functions encourages use of numeric and graphic representations that provide insight into how specific points of interest fit into the overall relationships of variables. For example, as one examines the graph and table of values for the profit function in a neighborhood around the maximum point, it becomes clear that moderate changes in ticket price will have very little effect on event profit.
- The concept of function is central to calculus. So having students encounter functions in their introduction to algebra lays important conceptual foundation for later studies.
- Functions in algebra connect naturally and effectively to transformations in geometry that students will use to reason about congruence and similarity.
- Statistical methods for data analysis and modeling lead naturally to functions as representations of relationships between correlated variables.
- Iteratively defined functions play a fundamental role in many applications of discrete mathematics to questions in finance and population dynamics.

Algebraic notation is valuable for representing what we know or what we want to find out. Algebraic procedures for manipulating symbolic expressions and equations into alternative equivalent forms are useful for gaining insight into relationships between variables. But there are now many powerful tools for doing that work. So algebra courses focused on developing skill in formal symbol manipulation are a poor use of valuable instructional time for all but a few students.

7.2 A Conceptual Framework for School Algebra

A more useful conceptual framework for thinking about school algebra can be expressed with a diagram that has become common in discussions of the mathematics curriculum. As students explore a numeric pattern or problem, they find ways to represent relationships between variables. They use these representations to reason about the situation in a variety of ways—solving equations and inequalities, understanding relationships, making predictions, and verifying patterns. Then when their mathematical model has suggested new insights into the problem situation, those ideas have to be evaluated to see how they play out back in the real world.

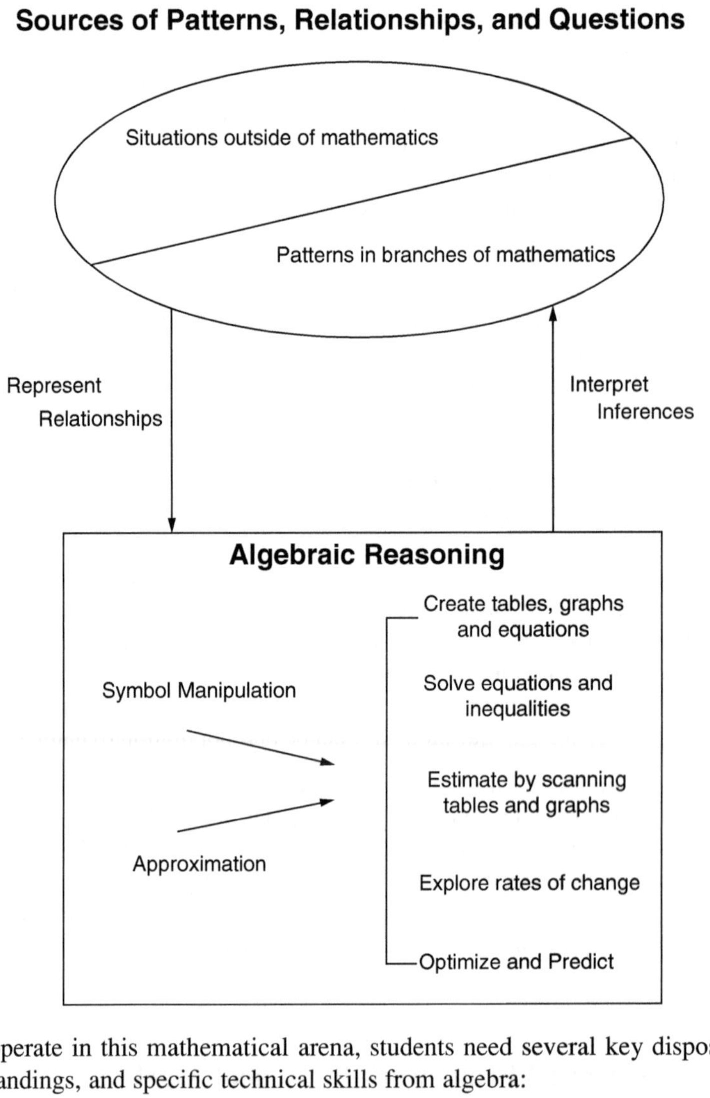

To operate in this mathematical arena, students need several key dispositions, understandings, and specific technical skills from algebra:

- **Disposition to look** for quantitative variables in problem situations and for relationships among variables that reflect *cause-and-effect*, *change over time*, or *pure number* patterns.
- A **repertoire** of significant and common patterns to look for—linear, quadratic, exponential, inverse variation, and periodic functions.
- Ability to **represent** relationships between variables in words, graphs, data tables and plots, and in appropriate symbolic expressions.

- Ability to **draw inferences** from represented relationships by estimation from tables and graphs, by exact reasoning using symbolic manipulations, and by insightful interpretation of symbolic forms.
- **Disposition to interpret** mathematical deductions in the original problem situations, with sensitivity to limitations of the modeling process.

These goals suggest a presentation of school algebra that begins by drawing students' attention to the many interesting situations in the worlds of science, business, engineering, and technology where quantities change naturally over time or in response to changes in other related quantities. The symbolic notation of algebra can be introduced naturally to make precise and efficient representations of observed patterns. Then students can learn, with a very modest amount of introductory personal symbolic reasoning, how to use the widely available array of computing tools to answer questions about the observed situations. For those students who ultimately need sophisticated and efficient personal skills for symbolic work and understanding of algebra that includes the formal structural aspects of the subject, we are now in a position to provide personal skill development when it appears essential, rather than as the first step toward proficiency in algebra-assisted reasoning.

In some sense this way of thinking about school algebra turns the traditional sequence of mathematical ideas, skills, and applications upside down—developing concepts and problem solving before personal symbol manipulation skills. But, in addition to providing broadly useful mathematical understandings and technology-assisted skills, the function-oriented development provides students with intuitions about variables, expressions, and equations that are a very effective concrete grounding for later development of the formal aspects of algebra. The syntactic rules of symbolic algebra become procedures that just make sense, rather than formal logical consequences of abstract field axioms.

7.3 A Sample Function-Oriented Curriculum

The curriculum projects mentioned above (IMP and CPMP) show how the proposed development of algebra in the context of functions can be accomplished. For example, the algebra/functions strand in Core-Plus Mathematics includes 15 units over the course of four high school years, units that are woven together with topics in other content strands. Each unit develops fundamental understandings and skills in use of functions and algebraic reasoning to solve problems in mathematics and its applications to science, business, and everyday life.

- *Patterns of Change* focuses on quantitative variables using data tables, coordinate graphs, and symbolic expressions.
- *Linear Functions* focuses on relationships between variables characterized by constant additive rate of change, straight line graphs, and equations in the general form $y = mx + b$.

- *Exponential Functions* focuses on growth and decay patterns characterized by constant multiplicative rate of change and expressed by the general form $y = Ab^x$.
- *Quadratic Functions* focuses on relations between variables expressed by the general form $y = ax^2 + bx + c$.
- *Functions, Equations, and Systems* focuses on relationships between two or more variables that can be expressed as inverse variations $y = \frac{k}{x}$, power functions $y = kx^r$, and systems of linear equations with two independent variables.
- *Matrix Methods* develops concepts and operations on matrices to represent and solve multivariable problems in algebra, geometry, and discrete mathematics.
- *Nonlinear Functions and Equations* introduces and develops formal symbolic methods for reasoning about quadratic functions, expressions, and equations, as well as logarithms for reasoning about exponential equations.
- *Inequalities and Linear Programming* focuses on algebraic and graphical reasoning about linear inequalities and systems.
- *Polynomial and Rational Functions* develops familiar concepts and skills in work with polynomials and rational expressions in the context of functions and their graphs.
- *Recursion and Iteration* develops properties of sequences as iteratively defined discrete functions, with special attention to arithmetic and geometric sequences and their connections to linear and exponential functions.
- *Inverse Functions* develops the inverse concept with special attention to logarithms and inverse trigonometric functions.[4]
- *Families of Functions* reviews and integrates student understanding of core function types and their representation in symbols, data tables, and graphs with a focus on transformation of basic function forms to model complex scientific relationships.
- *Algebraic Functions and Equations* develops core results in theory of equations and work with rational functions and equations.
- *Exponential Functions and Data Modeling* extends prior work with exponential and logarithmic functions and their expressions to the case of natural exponential and logarithmic functions, including use of logarithms for data linearization and modeling of patterns.
- *Concepts of Calculus* builds on prior work with functions, graphs, and rates of change to introduce core understandings about derivatives and integrals and their most common applications.

Note that the earlier units in this sequence don't attempt to teach students everything we know about any one topic. Each topic is revisited as necessary in later units and often in units that are not part of the algebra/functions strand.

[4] The trigonometric functions are developed in an earlier geometry/trigonometry unit titled *Circles and Circular Functions*. This is one example of the integration of strands in Core-Plus Mathematics that are roughly categorized as algebra/functions, geometry/trigonometry, statistics/probability, and discrete mathematics.

7.4 Challenges to the Function-Oriented Algebra Proposal

Teachers and mathematicians reacting to the proposed function-oriented view of school algebra raise a number of plausible questions about the approach.

Challenge: This is not really *algebra*. So much of the traditional content of algebra courses (such as factoring, expanding, and simplifying expressions, and solving equations) seems to be omitted or at least moved to the background.

Response: Whether a curriculum that highlights functions and moves formal symbol manipulation to the background is or is not *algebra*, is not the core question for consideration in school mathematics. The heart of the matter is whether functions make more sense as the mathematical spine of a secondary school curriculum than the long-standing approach that emphasizes formal manipulation of abstract expressions, equations, and inequalities.

Challenge: There are many mathematical problems and reasonings that are not well served by the focus on functions.

Response: While we can all imagine some interesting mathematical and applied problems that use facets of algebra not naturally developed through a focus on functions, we think functions and mathematical modeling are the place to start with most secondary school students. Furthermore, nothing proposed in the function-centric development rules out training students in more standard algebraic principles and skills as an extension of the focus on functions.

Challenge: Even the impressive capabilities of computer algebra systems lack essential symbolic flexibility capabilities like those that well-developed personal symbol manipulation skills can provide.

Response: It is certainly true that current computer algebra systems do not include the kind of subtle mathematician's intuition that can tell which equivalent form of a symbolic expression might be most useful in answering a specific algebraic question, nor do they have the flexibility to make nuanced variations on standard options. So relying on CAS for core symbol manipulation tasks places some inherent limitations on student algebraic reasoning performance. The standard response to this challenge is to argue, "Since we don't know which students will need highly developed symbol sense and skill, we should aim high for all students." However, as with all inclusion/exclusion decisions of curriculum design, there is an important cost-benefit calculation to be made. Is the time required to develop admittedly desirable symbol manipulation skill and intuition really time well spent? Evidence from long experience with algebra teaching suggests that the answer for most students is, "Probably not."

Challenge: The concepts-before-skills developmental sequence is not an effective learning trajectory—procedural skill takes a long time to develop, and one learns best by acquiring procedural skills and then having the structure of that skill domain become clear at a later point.

Response: 25 years ago there was little evidence that a change in priorities and developmental approaches to emphasize functions first and foremost would work with real students and teachers. But the intervening years have yielded a great deal of existence-proof evidence that those ideas are not so far-fetched. Furthermore, the power, access, and ease of use of calculating and computing tools have increased in dramatic ways from the days of the first graphing calculators and personal computers, and this trajectory seems certain to only accelerate in the near future. Thus if we aim to provide the kind of mathematical understandings and skills that will be useful and attractive to most students, a development of algebra that emphasizes functions and their applications can make a very strong claim for priority in school mathematics.

Challenge: Finally, professionals with knowledge of the history of mathematics education can argue fairly that proposals for integrated curricula and emphasis on functions have been around for over a century, but they never seem to take hold in practice.

Response: Even casual reading in the history of mathematics education reveals recommendations from many individuals and professional advisory groups to integrate topics in the high school curriculum[5] and to emphasize function as a central unifying idea.[6] However, neither recommendation had much impact on conventions in American mathematics education. The broadening of mathematical sciences during the twentieth century, especially the growth of probability, statistics, and computer science, makes the case for a broader and more unified school curriculum with new urgency. With respect to the recommendations about centrality of functions, we argue that the emergence of digital technologies, especially graphing calculators and computer algebra systems, has changed conditions for mathematics education in ways that make teaching about functions more natural and effective than ever before. As indicated in the ticket-price example, access to graphing tools makes it natural and insightful to look for solutions of equations such as $5000x - 65x^2 = 0$ by scanning the graph of $I(x) = 5000x - 65x^2$ for x-intercepts. Use of computer algebra systems to find exact solutions (and other complex symbol manipulations) should be quite appropriate skill for most students.

[5] For example, in his famous 1902 retiring presidential address to the American Mathematical Society, E. H. Moore urged schools to "abolish the 'watertight compartments' in which algebra, geometry, and physics were taught." Similar recommendations appeared in the 1912 *Report of the American Commissioners of the International Commission on the Teaching of Mathematics*, the 1923 Mathematical Association of American National Committee on Mathematical Requirements' *The Reorganization of Mathematics in Secondary Education* (Jones & Coxford Jr, 1970; Kilpatrick & Izsak, 2008).

[6] Emphasis on functions and interrelationships within mathematics had been made as early as the middle of the nineteenth century by the distinguished German mathematician, Felix Klein. That same thematic recommendation was picked up by curriculum advisory reports in the United States throughout the twentieth century.

7.5 Summary

We believe that development of important algebraic concepts and techniques through an approach emphasizing functions offers very attractive opportunities to provide powerful mathematical understandings and skills as part of an integrated curriculum. The necessary tools and textbooks and teaching strategies all exist, and we owe it to students of the twenty-first century to see that they are adequately equipped for the world in which they live.

References

Benjamin, H. R. W. (1939), pseudonym J. Abner Peddiwell. *The Saber-Tooth curriculum.* New York: McGraw-Hill. Republished by McGraw-Hill as *The Saber-Tooth curriculum, classic edition*, 2004.

Common Core State Standards for Mathematics. Washington, DC: National Governors Association Center for Best Practices and the Council of Chief State School Officers. Retrieved from http://www.corestandards.org/Math

Fendel, D., Resek, D., Alper, L., & Fraser, S. (2015). *Interactive mathematics program 1, 2, 3, 4.* Mount Kisco, NY: It's About Time.

Hirsch, C., Fey, J., Schoen, H., Hart, E., & Watkins, A. (2014). *Core-plus mathematics I, II, III, IV.* Columbus, OH: McGraw Hill.

Jones, P. S., & Coxford, A. F., Jr. (Eds.). (1970). *A history of mathematics education in the United States and Canada.* Thirty Second Yearbook of the National Council of Teachers of Mathematics. Washington, DC: National Council of Teachers of Mathematics.

Kilpatrick, J., & Izsak, A. (2008). A history of algebra in the school curriculum. In C. Greenes & R. Rubenstein (Eds.), *Algebra and algebraic thinking in school mathematics.* 2008 Yearbook of the National Council of Teachers of Mathematics. Reston, VA: Author.

National Council of Teachers of Mathematics. (1989). *Curriculum and evaluation standards for mathematics teaching.* Reston, VA: Author.

National Council of Teachers of Mathematics. (2000). *Principles and standards for school mathematics.* Reston, VA: Author.

Rehmeyer, J. (2014). Transitions: Organizing principles for algebra curricula—The transition from arithmetic to algebra. In *Critical issues in mathematics education series teaching and learning algebra workshop 5 2008.* Berkeley, CA: Mathematical Sciences Research Institute.

Chapter 8
Teaching and Learning Middle School Algebra: Valuable Lessons from the History of Mathematics

Mala S. Nataraj and Mike Thomas

Abstract Algebra is often thought of as a 'gatekeeper' in school mathematics, being crucial to further study in mathematics as well as to future educational and employment opportunities. However, a large number of studies have highlighted the difficulties and cognitive obstacles that students face when they learn algebra. In response to growing concerns about students' fragile understandings and preparation in algebra, recent research and reform efforts in mathematics education have made algebra curriculum and teaching a focus of attention. Very little research, however, has paid attention to extracting ideas from the history of algebra for developing classroom teaching strategies. In this chapter, we examine some important issues in the history of algebraic ideas involving *variables* and *exponents* that can transfer well to the mathematics classroom of today.

Keywords Algebra • Learning • Variables • History of mathematics • Pedagogy • Exponentiation

Four different perspectives on school algebra have been described (Usiskin, 1988) as: (1) algebra as generalised arithmetic; (2) algebra as problem-solving; (3) algebra as the study of relationships among quantities; and (4) algebra as the study of structures. Fundamental to these conceptions are the ideas of *variable*, *powers involving variables*, generalisation of patterns, and forming and solving equations. This chapter will present some key issues from the history of algebra and analyse how they may be used to enhance middle school students' algebraic thinking and reasoning. Some questions that will be addressed include: When and how did the idea of different variables develop? What were the generalisations made? What was the notation for powers? What are the implications of these historical developments

M.S. Nataraj (✉)
Selwyn College, Auckland, New Zealand
e-mail: natarajm@selwyn.school.nz

M.O.J. Thomas
The University of Auckland, Auckland, New Zealand
e-mail: moj.thomas@auckland.ac.nz

S. Stewart (ed.), *And the Rest is Just Algebra*, DOI 10.1007/978-3-319-45053-7_8

for modern day teaching? Based on the findings from the historical (and psychological) analysis, descriptions for teaching sequences and their implementation in middle or lower secondary school (age 13–14 years) are presented. The effects of this on student algebraic thinking suggest a vital role for lessons from history in the teaching of school algebra. It is to be noted that while the student results are from a school in New Zealand, what is discussed in this chapter is probably valid in most countries including the USA, given that students' difficulties in algebra is a worldwide phenomenon, as documented in the research literature.

8.1 Learning Algebra

Algebra learning in middle and secondary school often comprises pattern generalisation, use of unknowns, variables and powers of variables, forming expressions and equations, and solving equations. Students are immediately aware that algebra involves letters but there is clear evidence in documented research that many of them have very little grasp of what the letters mean and the reason that they are used (e.g. Graham & Thomas, 2000; Kieran, 1992; Küchemann, 1981). Students also have difficulties with algebraic symbolism, including exponential notation (Lee & Wheeler, 1989; MacGregor & Stacey, 1997; Pitta-Pantazi, Christou, & Zachariades, 2007), and exhibit a lack of understanding of basic mathematical terms such as 'solve' and 'evaluate' (McGowen, this volume). Students' difficulties in algebra are not surprising; Kieran (this volume) discusses two studies in the area of cognitive neuroscience and mathematics education and suggests that the findings are contrary to the traditional belief that it is 'just algebra' (meaning that algebraic problem-solving is simply a mindless execution of an automatic set of procedures). She further concludes that: (1) the algebraic symbolic method of solution is more demanding than a model diagrammatic method and (2) greater cognitive effort is required for achieving excellence in algebra. That is, the conclusions in the studies challenge the thought that students who excel are naturals at it. In this section we will consider some common misconceptions and difficulties faced by students related to variables and exponentiation. Awareness of these obstacles will be helpful in constructing pedagogical strategies to alleviate students' problems and to enhance understanding.

8.2 Variable

Algebraic thinking and reasoning depends on an understanding of key ideas, of which variable is one of the most fundamental. A large number of research studies have documented student difficulties related to the idea of the variable. As Schoenfeld and Arcavi (1988) point out, the variable concept is problematic for students because it is used with different meanings (for example, specific unknown, generalised number, variable, parameter and constant) in different situations.

Not only that, but students have also to grapple with the meaning of two or more different variables within an expression, such as x, y and z in $4x^3y$ and $12x^2z^5$. However, these different interpretations of the literal symbols are crucial in developing algebraic thinking and in the transition from arithmetic to algebraic thinking.

In the study conducted by Küchemann, and confirmed by Malisani and Spagnolo (2009) in their investigation of the role of variable in students' algebraic thought, the predominant conception seemed to be that of the unknown (when students did progress from viewing letters as numerical placeholders) and the variable as representing a range of values was beyond most students' grasp. In Küchemann's study, students were asked 'which is larger, $2n$ or $n + 2$?' and 70 % of the students gave the answer $2n$, giving reasons such as 'because it's multiply'. In order to answer this question correctly, students need to see n as a variable, to appreciate that n can take range of values. Then they might see that $2n$ is only greater than $n + 2$ when $n > 2$. In addition, many students appear to think that different letters in an expression will always take different values; that they can never be the same. In Kuchemann's study, for the question 'Is $L + M + N = L + P + N$ true always, sometimes or never?' 51 % of the students (13–15-year-olds) responded with 'never'.

The above response may stem from a static view of an expression such as $3x + 4y$ where x and y can take one set of different values rather than the dynamic view of x and y representing *differently changing values* that can sometimes be the same. In this connection, Küchemann (1981) states that the construction of a formula or rule seems to be built upon the notion of generalised number. In view of this, Radford (1996) contends that the idea of generalised number is a pre-concept to that of variable, and puts forward the view that the ways of thinking associated with generalisation (involving generalised number, variable and parameter) and equation solving (involving specific unknowns) are 'independent and essentially irreducible, structured forms of algebraic thinking' (Radford, 1996, p. 111). The observation that the generalisation and equation solving approaches seem to be mutually complementary domains in algebra instruction would imply that, at the foundation level, students need to grasp both unknown and general/ generalised number in order to make progress in algebra. Understanding the literal symbol as some kind of generalised number which can take a *range of values* is seen by some educators to provide a channel from a view of letter as unknown to that of letter as variable (Bednarz, Kieran, & Lee, 1996; Knuth, Alibali, McNeil, Weinberg, & Stephens, 2005). In addition, students need to also make sense of *different* variables (and *powers* involving variables) and parameters (or givens) in algebra. Hence, the importance of student activity focussed on generalisation is widely acknowledged in the research literature on the teaching and learning of algebra (e.g. Mason, 1996; Mason, Graham, & Johnston-Wilder, 2005). Indeed, Mason has emphasised the paramount importance of enabling students to express generality in every mathematics lesson. He states:

> Generalisation is the heartbeat of mathematics and appears in many forms...the heart of teaching mathematics is the awakening of pupil sensitivity to the nature of mathematical

generalisation and dually, to specialisation; that algebra as it is understood in school is the language for expression and manipulation of generalities. (Mason, 1996, p. 65).

…. *a lesson without the opportunity for learners to express a generality is not in fact a mathematics lesson.* (Mason et al., 2005, p. 297). (italics by authors)

As well as variables, students have to also grapple with expressions and equations in algebra that include exponents, and this is discussed next.

8.3 Exponents

Exponents are important mathematical concepts and central to many secondary and tertiary mathematics courses including algebra, calculus and complex analysis. For example,

1. Very large numbers (and very small numbers) which are ubiquitous in today's society are able to be expressed using exponents, including scientific notation.
2. The concept of place value is based on exponentiation and hence a good grasp of exponents leads to a deep understanding of positional notation.
3. Algebraic expressions and equations often involve exponents.
4. Proficiency in the manipulation of algebraic expressions including exponents is essential for students who want to enter STEM courses in tertiary education.

Even at the high school level, in New Zealand, more than half the questions in the Year 11 (age 16 years) algebra exam (*Mathematics Common Assessment Task* or MCAT) (NZQA, 2014) consisted of exponents in some form or the other, and in the Year 12 algebra exam, *all* the questions involved some understanding of exponentiation. In New Zealand, students complete schooling at Year 13, and mathematics is a compulsory subject up to Year 11. Despite its importance, there have been relatively few studies focused on students' understanding of exponentiation. The studies that do involve exponents have highlighted some errors that students are prone to make, such as writing x^3 instead of $3x$ (MacGregor & Stacey, 1997), assuming that n^2 and 2^n are the same (McGowen, this volume) and expanding $\left(a^2 + b^2\right)^3$ to $a^6 + b^6$ (Lee & Wheeler, 1989). Hence, even at the 'action' stage of APOS (action, process, object and schema) theory (Arnon et al., 2014; Dubinsky & McDonald, 2001), students experience problems. That the negative sign and rational exponents is a problem for students is underscored in the study by Pitta-Pantazi et al. (2007) and Fig. 8.1 shows the success rates for 5 out of the 20 tasks given by them to 202 high school students.

Other errors that have been reported relate to operations on powers involving positive integer exponents such as (1) $2^2 \times 2^3 = 2^6$, (2) $5^6 \div 5^2 = 5^3$, (3) $5^2 - 3^2 = 2^0$ and (4) $2^3 \times 3^3 = 6^6$ or 6^9 (MARS, Shell Centre, 2015). Similar errors were noticed by Cangelosi, Madrid, Cooper, Olson, and Hartter (2013) when they administered a comparable assessment to 904 freshman and sophomore university students. The researchers sought to identify persistent errors that students make when simplifying

Compare the exponents using the symbols <, =, or >	Low Achievers' performance (%) (Group 1)	Average Achievers' performance (%) (Group 2)	High Achievers' performance (%) (Group 3)
1. 23^{8}, ….. , 23^{13}	97.6	97.4	99.2
2. 23^{-8}, ….., 23^{-13}	31.0	83.6	94.7
3. $(-12)^{13}$, ….. , $(-12)^{17}$	33.3	75.4	97.4
4. $(-12)^{-7}$, ….. , $(-12)^{-9}$	39.3	55.3	66.7
5. $17^{(-3/5)}$, ….. , $15^{(-3/5)}$	21.4	59.8	84.2

Fig. 8.1 High school students' success rates in tasks related to exponents (from Pitta-Pantazi et al., 2007)

exponential expressions and to understand why students make these specific errors. Among other mistakes, some of the errors that students made were:

1. Simplifying -3^2 to 9
2. Writing $-9^{3/2}$ as $(-9)^{3/2}$
3. Simplifying 2^{-3} as $2^{1/3}$ or -2^3 or 2/1/3 or 3/2

The authors noted that a few students used the definition $a^{-1} = \frac{1}{a}$ to remind themselves of a way of simplifying 2^{-3}. Looking at the symbolisation, they wrongly generalised the definition to the incorrect statement $a^{-2} = \frac{2}{a}$, concluding that the base forms the denominator and the exponent forms the numerator. Hence, applying this erroneous generalisation, some students wrote $2^{-3} = \frac{3}{2}$. The authors (Cangelosi et al., 2013) concluded that an inadequate understanding of the notion of negativity as the source of most of the students' errors in their study, and conjectured that a deeper understanding of additive and multiplicative inverses could alleviate the problem and aid students to develop a more abstract view of negativity. They also suggest that the effect of language, notation and grouping could be factors contributing to student misconceptions. If students have difficulties with *exponential numerals* then it is not surprising that they struggle with expressions involving *powers involving variables* such as k^5, 7^m, x^y and $8p^3q^6$. In this context, Weber (2002a, 2002b) examined post-secondary students' ideas about expressions involving exponents in the context of APOS theory. He suggests that the main difficulty for most students is that while they are able to grasp exponentiation as an action, many did not understand this concept as a process, which is necessary to comprehend *exponentiation* as a *function*, with its full range of real number values. In turn, the ideas involving exponentiation as function play a pivotal role in students' deep understanding of calculus and advanced mathematics. As seen above, many students have only a fragile understanding, even at the 'action' stage of exponentiation.

What is implied above is the need to provide students with more experiences in the interpretation of exponential forms. If problematic issues are not addressed in time, or if students' prior experiences are not wide-ranging, this can lead to long-term consequences that hinder learning. Tall (2004) coined the term *met-before* for key ideas that students have met before that may influence their current understanding, both positively and negatively. Hence, he suggests it is necessary to pay close attention to how new learning is affected by prior experiences of students and how these may support or hamper acquisition of new knowledge. Earlier learning can be *supportive* on occasions where old ideas can be used to understand new content, and *problematic* where the old understanding no longer makes sense in the new situation. For example, in this book, McGowen discusses the minus sign as a problematic met-before. One implication of these ideas is the need for students to be exposed to opportunities to develop a wide conceptual base in the early and middle grades comprising supportive met-befores. These supportive met-befores may form a springboard for the assimilation of new ideas that students will encounter in the future at high school and at university.

The foregoing has highlighted how research describes students' inadequate understanding of the variable concept and its notation, and their limited grasp of exponential numerals and expressions. In their learning of algebra, students have to make sense of, and to manipulate different variables and powers of variables in order to solve problems and to progress to advanced mathematics. In an attempt to alleviate this situation and to develop alternative/supportive approaches, we turn to the history of mathematics.

8.4 History of Mathematics and Mathematics Education

In view of the difficulties that students face in understanding mathematics, for some years now, educators and researchers have consulted the history of mathematics in attempts to improve the teaching and learning of the subject (Fauvel & van Maanen, 2000). One of the emerging aspects of such studies concerns investigations of the historical development of mathematical ideas. The theoretical framework sometimes employed in historical inquiry is that *ontogenesis* recapitulates *phylogenesis*. That is, the mathematical development of the individual student repeats that in the history of mathematics (Radford, 2000; Sfard, 1995). In Piaget and Garcia (1989), the claim is made that reconstruction of the history of science cannot be separated from a psychological analysis, and from a Vygotsky (1978) perspective, historical review is vital to establish what concepts it is most important to teach, and in what order. While a strict historical parallelism is usually considered indefensible, two possible didactic uses of history are first, to understand better the cognitive difficulties experienced by our students, and second to make more enlightened decisions concerning the knowledge being taught in the classroom. Taking into account the order in which past conceptual developments occurred, a teacher may use their knowledge of history to design classroom activities for enhancing the understanding of mathematics in which historical ideas may be present explicitly or implicitly.

We present here some examples of the kinds of studies in mathematics education that have involved a historical-critical analysis related to understanding in algebra:

1. In her work on negative numbers and elementary algebra, Gallardo (2001, 2008) reviewed a Chinese text, a medieval Italian text and a treatise from the nineteenth century and investigated the problems of learning and teaching negative numbers and elementary algebra.
2. A comparative analysis of the history of algebra with students' empirical data by Harper (1987) found a parallelism between the evolution of algebraic symbolism (the rhetorical, syncopated and symbolic stages) proposed by Nesselman in 1942 and the way students understand letters in algebra. Harper foregrounded the fact that it took a long time in history to make the shift from syncopated to symbolic algebra (including parameters or givens) and advocated the use of history in understanding students' difficulties and to explicitly make students aware of the different usage of letters. Similarly, Sfard (1995) maintains that the history of mathematics is indispensable to make teachers and educators alert to deeply hidden obstacles concerned with new concepts such as variables and parameters.
3. Given the historical evidence that the interest in constructing general methods for solving sets of similar problems was the basis for the development of algebra, Ursini (2001) devised a pre-algebraic experience for primary pupils using Logo. Ursini used students' numeric background as a support for different uses of the variable.
4. In her thesis, van Amerom (2002) enquired into the teaching-learning process pertaining to the transition from arithmetical to algebraic problem-solving by drawing on the historical development of algebra.
5. Following a historical review, Schmittau (1993) reports a study with university students on the conception of exponentiation, beginning with the exponential function. This approach, Schmittau suggests, allows for the full range of real number exponents to emerge from the attempt to solve a problem in which it is required to express mathematically (both graphically and as a function), the situation of continuous growth.
6. Subramaniam and Banerjee (2011) review a discussion of the relation between arithmetic and algebra in an Indian historical text from the twelfth century and conclude that algebra is more a matter of understanding and insight than the employment of symbols and that algebra is seen as foundational to arithmetic rather than as a generalisation of arithmetic. The authors present a framework that highlights the arithmetic–algebra link and report briefly on a teaching approach that is informed by this framework.
7. From a different perspective to the above studies, Katz (2007) has argued that besides the three stages of expression (rhetorical, syncopated and symbolic), four conceptual stages have happened alongside the notational stages; these are the *geometric* stage, the static *equation-solving* stage, the dynamic *function* stage and finally the *abstract* stage. This would imply that since most algebra concepts were represented geometrically in the initial stages in the history of mathematics, the use of geometry as a tool maybe a useful pedagogical and curricular

approach in the teaching of algebra as demonstrated by both Tall (for $a^2 - b^2$ and $a^3 - b^3$ in this volume) and Mason (in this volume).

Thus far we have looked at students' difficulties, and the importance of understanding the concepts of, generalised numbers, variables, powers of variables, and the use of this notation in solving problems. We have also glimpsed some ways in which the history of mathematics has been incorporated in teaching and learning. In the following section we place the use of variables and their powers, in a context of the development of this symbolism in the history of mathematics, with a view to incorporating ideas into teaching to enhance understanding of algebra.

8.5 Learning from History: Variables in Algebra

A review of historical texts shows that for a long time all equations representing problem situations were written using only one letter. That is, only one unknown was used in an equation and other unknowns were represented in terms of this one unknown. Radford (1995) underscores the fact that the second unknown/variable came relatively late in history and hence it is not surprising that students find the idea challenging. An examination of Viete's (1540–1603) work discloses his use of different letters for different variables for the construction of expressions and equations. Viete reserved x, y and z for variables and a, b and c for parameters. Unlike his predecessors Diophantus and the Arab mathematicians, whose equations involved only a single unknown (Van Der Waerden, 1954), Viete was able to form and solve equations involving different unknowns/variables. However, nearly a 1000 years prior to Viete, the Indian mathematician Brahmagupta (598–670) had employed different letters for different variables (although not literal coefficients, the Indians had a *common name* for parameters), and had a terminology for powers of variables. A study of the development of algebraic symbolism during this period reveals some interesting ideas for the teaching and learning beginning algebra.

A historical analysis of Indian mathematics reveals that the current decimal numeration system with place value, zero and *distinct* symbols for the nine digits, as well as many algebraic ideas, originated in India (e.g. Bag & Sarma, 2003; Cajori, 1919; Datta & Singh, 2001; Eves, 1969; Joseph, 2011). For example, Puig and Rojano (2004) cite how a Mathematical Sign System (MSS), in which the *different unknown quantities* and *their powers* are differentiated, an important step in the development of algebraic notation, was constructed in India by the time of Bhaskara II in the twelfth century, or possibly even earlier in the time of Brahmagupta's in the seventh century (Colebrooke, 1817). However, this was achieved by Viete only in the sixteenth century in Europe. This construction, the chief characteristic of which is the notation of *different variables* and their *powers*, enabled the development of *general methods* of solutions of equations. One specifically interesting characteristic of the MSS developed in India was that various *colours* (and later their abbreviations) were used to denote *different unknowns*.

Bhaskara II (1150) says: '*yavat-tavat* (so much as), *kalaka* (black), *nilaka* (blue), *pitaka* (yellow), *lobita* (red) and other colours have been taken by the venerable professors as notations for the measures of the unknowns, for the purpose of calculating with them' (Datta & Singh, 2001, p. 18). Thus Bhaskara II employed abbreviations of the names of the unknown quantities in order to represent them in an equation, such as *ka* for *kalaka* (black) and *ni* for *nilaka* (blue). Although *yavat-tavat* (quoted above) is not a colour its inclusion shows the persistence of an ancient symbol employed long before colours were introduced to denote unknowns.

One implication of the historical development in India is that while students can be (and mostly are) acquainted with algebra through the solving of equations involving specific unknown(s), they may benefit from meeting generalised numbers and variables sooner than its present introduction into the curriculum. The need for generalisation as a way of thinking (Mason, 1996; Radford, 1996) suggests that at the fundamental level, students need to grasp both concepts and these have to be explicitly taught (Harper, 1987) in the classroom. For example, students need to understand the meaning of letter p both in expression '$3p + 2$' and in equation '$3p + 2 = 17$', namely letter as generalised/general number and letter as unknown. Instead, most students tend to view letters only as specific unknowns particularly due to their experiences involving substitution and equation solving, and hence show a lack of familiarity of letters representing a range of values.

In order to test the value of lessons from Indian development of mathematical thinking in algebra a teaching framework (see Fig. 8.2) based on these historical ideas and recent psychological literature was developed by Nataraj (2012).

The aim of the framework was to answer the question, how can generalised numbers be introduced to students? In this context, it was considered that the idea of colours (as signs) in Indian history to denote different unknowns could prove useful.

The need for improvement was clearly shown in the Concepts in Secondary Mathematics and Science (CSMS) study (Küchemann, 1981) involving high school students' interpretation of literal symbols. This demonstrated that a majority of the students—73 % of 13-year-olds, 59 % of 14-year-olds and 53 % of 15-year-olds—either treated letters as concrete objects or ignored them. In order to address this deficit of understanding a teaching module for beginning algebra students based on the abstract nature of colours (without reference to any particular object) and pattern language (see following paragraph and Fig. 8.3) was produced. Another beneficial feature of the use of colours in the module is that they stress the visual aspect in recognising changing numbers.

According to the theory of the structure of attention proposed by Mason (2004) we may focus our attention on the whole, the details, the relationships between the parts, the properties of the whole or the parts, or deductions, becoming more aware of what we notice. Mason also says (Mason et al., 2005) that classification is a form of generalisation and declares that children have a natural ability to classify objects. Thus in order to detect generalities in arithmetic patterns, he suggests guiding students' attention towards number patterns by asking questions such as 'what is the same about each row?', 'what is different and how is it changing'? This same

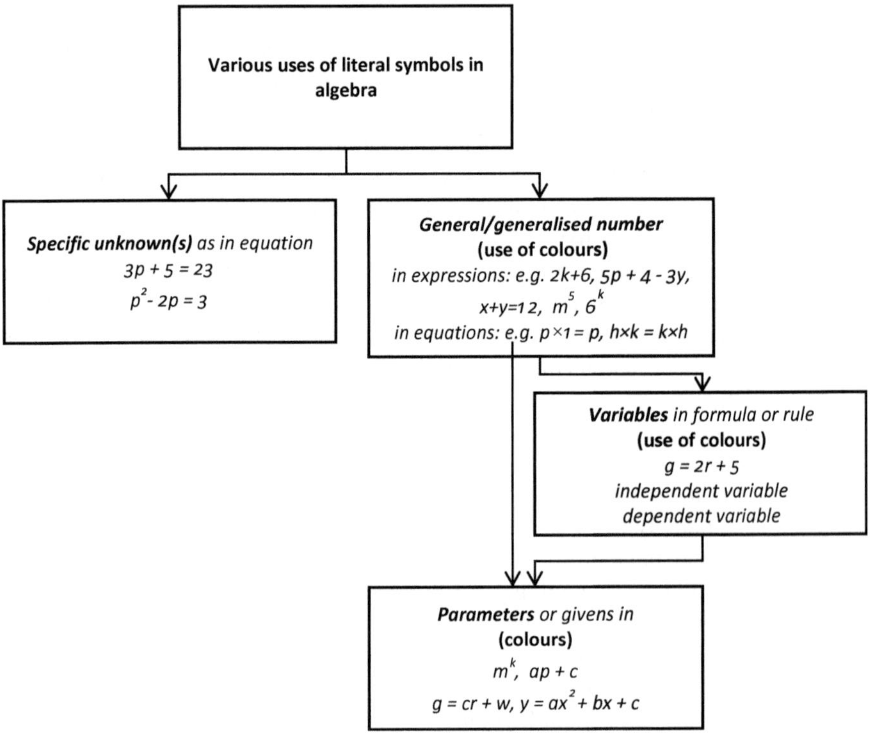

Fig. 8.2 Nataraj's (2012) framework for teaching the various uses of literal symbols in algebra

method of guiding attention was also described by Srinivasan (1989) who advocated the use of '*pattern language*' for number patterns and '*design language*' for shape arrangements to concentrate students' thinking on variation and invariants in number and geometrical patterns. His recommended vocabulary that centres around *changing, not changing, changing in the same way* and *changing in different ways* in order to elicit an algebraic expression from students in the form of pattern language, and this was used in the module developed (Nataraj, 2012) (n.b., Srinivasan recommends the use of a 'wriggly' line as a separator between number patterns and expressions, rather than the use of the usual line segment. His reason for this is that attention should be paid to the patterns and not to the operational outcome).

A pedagogical aim of the module was to assist students to acquire a deeper awareness of generalised number and then variable by *combining* ideas of *pattern language*, and *colours/signs* from Indian history within a number pattern generalisation activity. The manner in which this combination (historical and psychological) was employed is outlined in Fig. 8.3.

The important focus here also is that, while students may understand that two different colours/letters indicate two different sets of changing numbers, they also

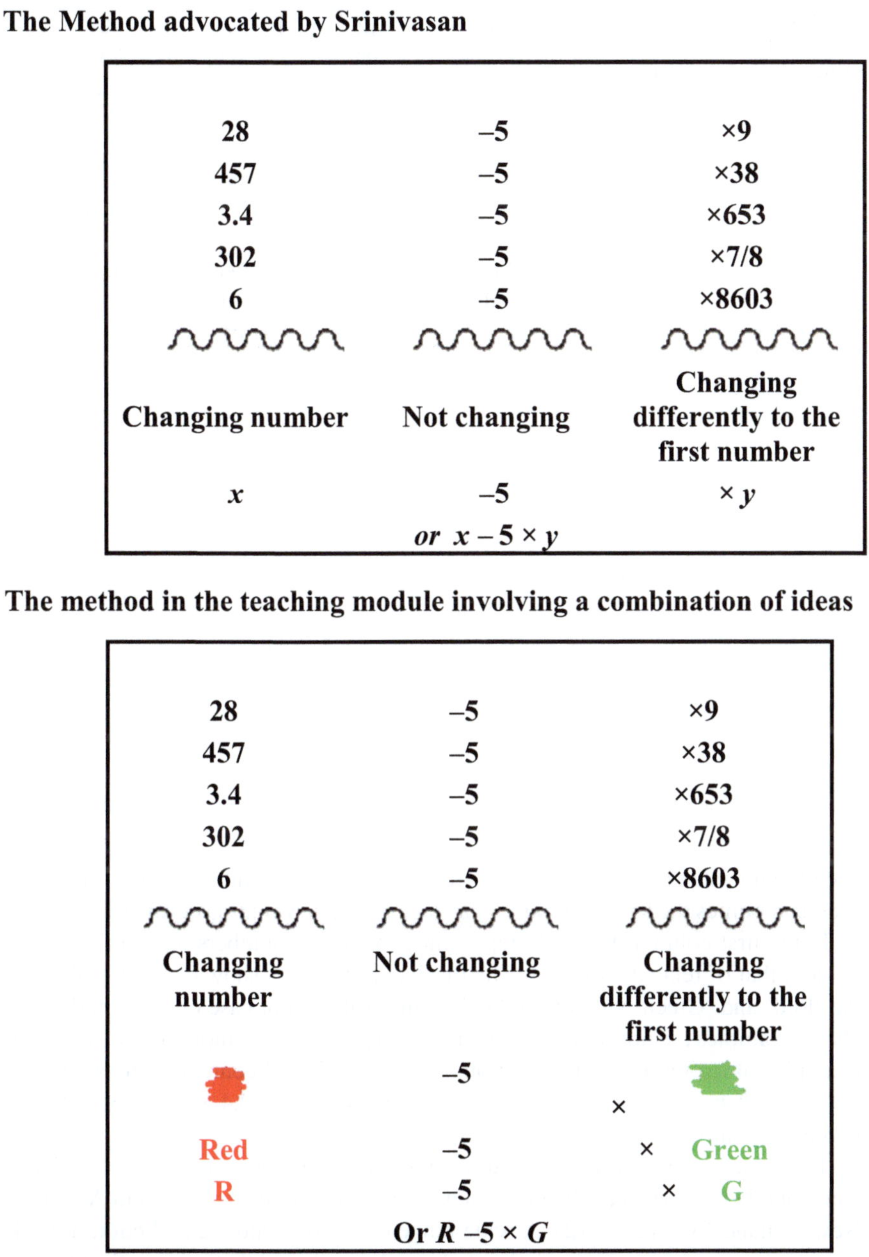

Fig. 8.3 Pattern generalisation using a combination of historical ideas and Srinivasan's method

Fig. 8.4 Pattern generalisation in an expression and in a formula

	a)			b)				
	4	+	12	7	+	6	=	13
	0.08	+	589	4.5	+	8.5	=	13
	6	+	6	1	+	12	=	13
	97	+	34.9	13	+	0	=	13
	15	+	15	6.5	+	6.5	=	13
	306.4	+	5^2	6	+	7	=	13
	4059	+	-7	5.64	+	7.36	=	13
	Red	+	Green	Red	+	Green	=	13
	R	+	G	R	+	G	=	13
				or		R = 13 –G		

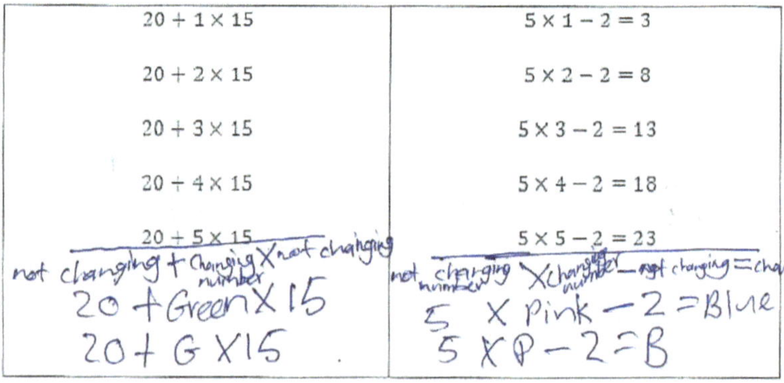

Fig. 8.5 Some of S26's answers showing the use of colours and letters to represent variables

need to understand that since we are choosing from an infinite number of possibilities, sometimes the letters/variables can take the same value, as outlined in Fig. 8.4.

In the first column, R and G represent generalised numbers and in the second column that of related variables. The third and fifth lines show examples of where the 'Red' and 'Green' variables take the same values. For ease of presentation, the above examples are not set in context, however, students are more likely to relate to examples of expressions and equations set in meaningful context such as '40 % of your pocket money + 30 % of your weekly earnings' (or $0.4 \times 25 + 0.3 \times 80$) and numerical variations of these.

These ideas were applied in a study involving 29 students (13 years of age) who were members of one Year 9 class in a multicultural secondary school in Auckland, New Zealand (Nataraj, 2012). A wide range of socio-economic and cultural backgrounds was represented among the students. It was found that prior to the teaching intervention only around 9 % of the students demonstrated the ability to recognise variation (partial understanding) or notate it using symbolic literals. In comparison, afterwards 80.2 % of the students displayed either partial or full understanding of varying quantities. In addition, 71 % of the students were able both to distinguish between variation and invariance and also symbolise the variation using a letter. Figure 8.5 shows an example of a student's work where they recognise the changing nature of some values and are able to use the colour construct above to go on to symbolise the expression. Some of the student comments following the intervention give some idea of the students' grasp of generalisation, variation and invariance.

S4: I think a generalisation is something like a letter or a colour that can be put in the place of something that can represent multiple things.
S6: I notice that there are three kinds of numbers—changing, not changing and changing but differently.
S7: I also understand the concept of generalisation is the number that is changing...to group a changing number with a letter or symbol.
S13: generalisation is when you find some numbers changing like 3^1, 3^2, you first choose any alphabet and put it as 3^b so you know those numbers are changing.
S18: what I understand about generalisation was that we put a symbol to show the numbers that change, and we leave and use the same number if it doesn't change.
S22: letters can take the value of any number.
S27: ....there had to be a letter that was the same because they represented that the numbers were changing the same.

In the case of parameter, students' inability to understand and use literal coefficients has been described in both Küchemann's (1981) and Harper's (1987) investigations, and affirmed by Sfard (1995) in her analysis of history of algebra and psychogenesis. However, in their study involving high school and college students' work with algebraic expressions and problems involving parameters, Ursini and Trigueros (2004) found that although students had difficulties in working with parameters, their difficulties decreased when a specific referent can be given to them. The authors recommend that parameters should be considered as generalised/ general numbers that are used to make second order generalisations.

Hence, combining the historical idea of colours along with a psychological perspective of the notion of parameters as second order generalisations, it was proposed (Nataraj, 2012) that the same didactic method described above can be extended (see Fig. 8.6) to assist students to give meaning to letter as parameter. In the first example, the specific referent is that of the straight line and $y = mp + c$ represents families of straight lines. Another possible approach, as outlined in the

Fig. 8.6 Second order pattern generalisation for understanding parameter

Parameter as second order generalisation	General method of solution as a tool for understanding parameters or givens
y=3p+42	3p=12, p=12/3
y=4p+16	4p=24, p=24/4
y=2p+15	2p=25 p=25/2
y=7p+21	4.75p=9.65, p=9.65/4.75
∿∿∿∿∿	∿∿∿∿∿
y=maroon×p+crimson or *y=mp+c*	*mp=v* *p=v/m*

second example, is that of the general method of solution (Ursini, 2001) as a tool for understanding parameter.

In summary, in order for students to be successful in algebra, they need to have a clear understanding of:

1. Letter as unknown ($3p + 2 = 14$)
2. Letter as generalised/general number ($3p + 2$)
3. Letters as variables ($3p + 2 = y$)
4. Letters as parameters ($mp + k = y$)

It is proposed that experiences such as the above could provide students with supportive met-befores (Tall, 2004) that will enable them to construct a conceptual base for a deeper understanding of algebra at senior school and at university.

8.6 Learning from History: Exponents

The discussion on student difficulties in understanding and working with exponents related above suggests that an examination of the historical development might assist with understanding student problems and provide ideas for the development of appropriate didactic strategies.

The current exponential notation was developed by Chuquet (fifteenth century) and Bombelli (sixteenth century) and the sign system of symbolic algebra that included powers of variables was fixed by the time of Euler in the eighteenth century (Puig & Rojano, 2004). However, this sign system for symbolic algebra was only achieved following a long history of naming and working with large numbers, that finally arrived at a system of notation that incorporated an adapted positional notation system into powers to denote exponents. Hence x^0 was a simple number or constant, then x^1 denoted x and $x \times x$ denoted x^2, and so on. A further study of history (e.g. Joseph, 2011) reveals that many cultures named and worked with large numbers, including the Egyptians and Mayans of Central and South America. For example, the Greek Archimedes, considered one of the greatest mathematicians, defined 'myriad' as 10,000 in his *Sand Reckoner*. Using the myriad, he expressed numbers up to a myriad-myriads which he called numbers of the first order. This in turn was the unit for the second order of numbers and so on to naming a number greater than the grains of sand that would fill the universe! This may be compared with the work of Indian mathematicians, whose traditional fascination with naming and working with large numbers allowed them to build a spectacular tower of numbers. What were the types of numbers that they considered in ancient times? A few examples from Indian historical texts are given below:

1. A major milestone in the development of the Hindu-Arabic place value system is a (surprisingly very early) set of number names for powers of ten. In the *Vajasaneyi* (*Sukla Yajurveda*) *Samhita* (17.2) (c. 2000 BC) of the Vedas, the following list of *arbitrary* number names is given in Sanskrit verse: *Eka* (1),

Dasa (10), *Sata* (10^2), *Sahasra* (10^3), *Ayuta* (10^4), *Niyuta* (10^5), *Prayuta* (10^6), *Arbuda* (10^7), *Nyarbuda* (10^8), *Samudra* (10^9), *Madhya* (10^{10}), *Anta* (10^{11}), *Parardha* (10^{12}) (e.g. Bag & Sarma, 2003; Datta & Singh, 2001). They were aptly called the *dasagunottara samjna* (decuple terms), confirming that there was a definite systematic mode of arrangement in the naming of numbers. The same list of names of powers of ten was then extended to *loka* (10^{19}) (Gupta, 1987).

2. In the Buddhist work *Lalitavistara* (c. 100 BCE), there are examples of series of number names based on the centesimal scale. For example, in a test, the mathematician Arjuna asks how the counting would go beyond *koti* (10^7) on the centesimal scale, and Bodhisattva (Gautama Buddha) replies: Hundred *kotis* are called *ayuta* (10^9), hundred *ayutas* is *niyuta* (10^{11}), hundred *niyutas* is *kankara* (10^{13}), . . .and so on to *sarvajna* (10^{49}), *vibhutangama* (10^{51}), *tallaksana* (10^{53}). It is to be noted that there are 23 names from *ayuta* to *tallaksana*. Then follow 8 more such series, starting with 10^{53} and leading to the truly enormous number $10^{53+8\times46} = 10^{421}$! (Menninger, 1969).
3. In the Vedic literature, time is reckoned in terms of *yugas* or time cycles. The four *yugas* are *Satya-yuga*, *Treta yuga*, *Dwapara yuga* and *Kali yuga*. According to Hindu cosmology, the time-span of these four yugas is said to be 1,728,000, 1,296,000, 864,000 and 432,000 years, respectively, in the ratio 4:3:2:1. The total of these four *yugas* was considered as one *yuga-cycle* or *Mahayuga* and was thus 4,320,000 years (Srinivasiengar, 1967). Moreover, it is believed that 1000 such *yuga-cycles* comprise one day in the life of *Brahma*, which is 4,320,000,000 years and one day and night period is 8.64 billion years which was further extended to 311×10^{12}. As pointed out by Plofker (2009) time in the astronomical works is bound by cosmological concepts. In one *kalpa* which is 4,320,000,000 years, all celestial objects are considered to complete *integer* number of revolutions about the earth.
4. Like the Vedic mathematicians, the Jaina mathematicians, as part of their philosophy took special interest in long stretches of time and space. One example says: Consider a trough whose diameter is that of the Earth. Fill it up with white mustard seeds counting one after another. Similarly fill up with mustard seeds other troughs of the sizes of the various lands and seas. Still the highest enumerable number has not been attained (Joseph, 2011).

What can be seen above is that very large numbers (leading to exponential numerals), some of which were related to time and distance, were considered in India and they held a special fascination for the ancients. Eventually, such considerations led to a place value numeration system based on exponentiation of powers of ten. Hence, students in the middle school may benefit from reading, writing and naming very large numbers and working with them. The generation of problems such as the 4 mentioned above have the potential to promote calculation and the development of *quantity sense* or a sense of the size of numbers (Wagner & Davis, 2010). In addition, such numbers were employed in a practical/realistic context (Plofker, 2009) such as astronomy, and time measures (for calendar purposes).

Students are also often fascinated by large numbers and may be even more motivated if a meaningful, authentic context is used (Jhagroo & Nataraj, 2015). Concepts involving exponents are usually taught at the secondary school level by abstract rules, which students accept and often with little or no visible authentic application. Textbooks often move from exponential numerals to powers involving letters within a few exercises on scientific notation. However, students' persistent misuse of exponents (MacGregor & Stacey, 1997) points to an insecure foundation of the concepts of multiplication, repeated addition and repeated multiplication. Even at the basic level of repeated multiplication, students' responses also reflect a lack of understanding of the nested effect of grouping of repeated multiplication such as 3 groups of 3 groups of 3 groups of 3 in $3 \times 3 \times 3 \times 3$, which is the same as 3^4. It seems that students need such grouping experiences (involving for example matchsticks), and work that leads to understanding of exponential concept for positive integers, in order to appreciate the difference between 3^4 and 3×4, and, 3^4 and 4^3, and so on. However, the definition of exponentiation as repeated multiplication alone presents other problems later on since students have to establish some sort of meaning to non-positive and non-integral exponents. In order to circumvent this, Schmittau (1993) suggests beginning the exponent concept with exponential functions and allowing for the full range of real number exponents in a problem where students are required to express mathematically a situation involving continuous growth.

Out of these considerations, Schmittau designed a teaching experiment that allowed for the emergence of non-positive and non-integer exponents (given in Fig. 8.7).

Rather than reflecting botanical reality, Schmittau says that this task, which she suggests introducing after powers with positive integers, entails movement between arithmetic and geometric sequences (Day and Height axes) through the development of the exponential function $y = 3^x$, and solving for x at various intervals, not only for positive integers, but also zero, negative, and fractional exponents are developed. Because of the continuous nature of plant growth, heights of the plants involving irrational exponents can be seen to be possible (see some potential results in Fig. 8.8).

Furthermore, powers provide a crucial link between arithmetic and algebraic notation and concepts and hence should be exploited in teaching and learning. For example, generalising place values (involving second order generalisation in two ways as shown in Fig. 8.9) could provide students with an opportunity to develop a depth of understanding not only of an idea as fundamental as positional numeration system but also a deeper awareness of key ideas in algebraic symbolism—different symbols being for different variables, and powers of variables.

In Fig. 8.9, the top half demonstrates horizontal first order generalisation and then vertical second order generalisation while the bottom half reverses this. Once again the historical review presented here, along with a psychological perspective on exponents, was used to design a teaching framework for implementation in the middle school and lower secondary school classroom, and this is given in Fig. 8.10. The teaching sequence has been constructed in order to enhance the conceptual

The Task: At 8.00 Sunday morning a child notices a small plant growing near his house. He decides to measure it and finds that it is 3cm high. He measures it again on Monday morning at 8.00 and finds it to be 9cm high. He decides to measure it at the same time on ensuing mornings. Tuesday's measurement is 27cm, and Wednesday's is 81 cm. Assuming that this growth pattern is descriptive of the entire growth history of the plant:

1. How tall was it on the previous Saturday morning at 8.00? Why didn't he notice it?
2. How tall was it the previous Friday at 8.00 am? The previous Thursday at 4 am?
3. If we label Sunday as Day 1, the first day the child measured the plant, and want to be consistent with our numbering scheme,how should we number the following days: Saturday?, Friday? Thursday? If we denote the height of the plant on Sunday at 8.00 am as 3^1 cm, how could we express the heights on the other days?
4. How tall was the plant at 8.00 the previous Saturday night? At 8.00 Sunday night? At 4.00 pm on Saturday?(Did you happen to find the height at another day or time?)
5. When will the plant be 46.765 cm tall?

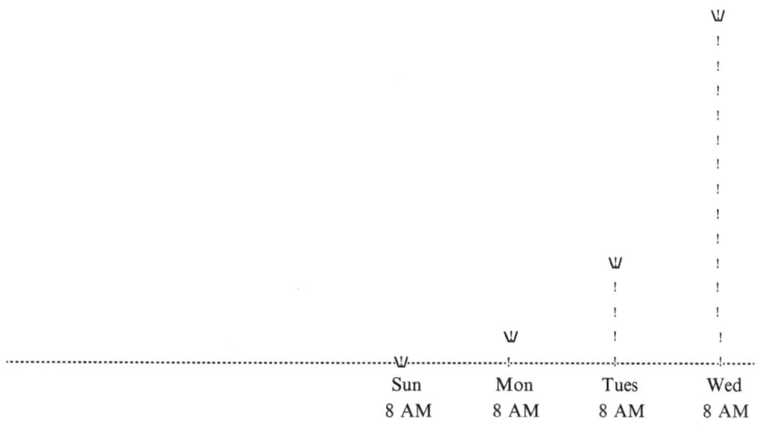

Fig. 8.7 The plant growth problem (Schmittau, 1993)

base that students hold for understanding exponentiation in arithmetic, and subsequently in algebra, and for progressing to advanced mathematics. This teaching framework was used with the same class of 29 Year 9 students described above.

Overall the students improved their test score after the intervention, with the $\text{Mean}_{\text{pre}} = 4.8$ and $\text{Mean}_{\text{post}} = 18.34$, $t = 10.87$, $p < 0.0001$. While there was no evidence of exponentiation beforehand, all apart from two students (one of whom had 5 weeks' absence) could symbolise powers afterwards. Their use of exponentiation in explaining place value may be seen in Fig. 8.11.

A few students were even able to generalise their use of exponentiation to general positions in a place value system with a general base, as seen in Fig. 8.12.

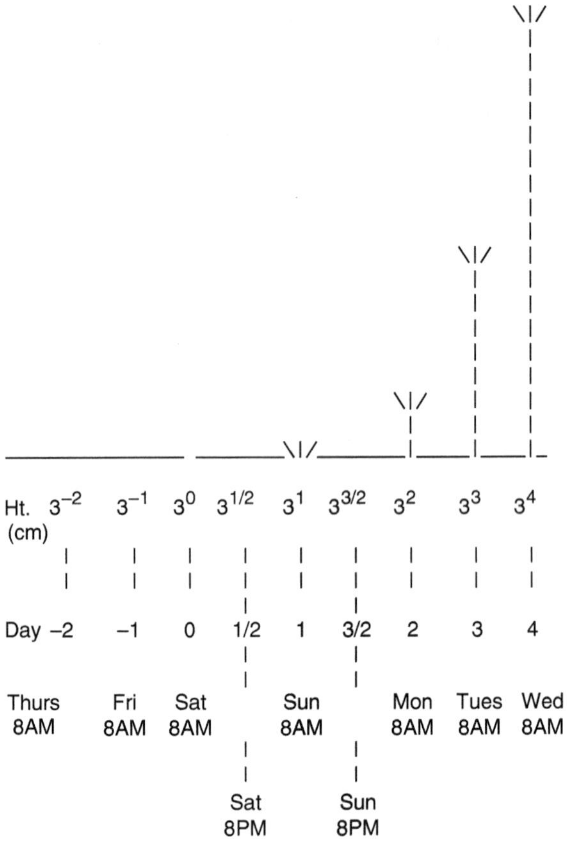

Fig. 8.8 Partial results of plant growth problem (Schmittau, 1993)

8.7 Concluding Remarks

The deliberations in this chapter cover the concepts of variable and exponentiation, along with students' understanding of these ideas, and demonstrate in practice the value of employing a historical perspective in teaching these ideas. However, it is also critical that students gain fluency in algebraic procedures (Hiebert & Lefevre, 1986; Skemp, 1976) and in harmony with this the National Council of Teachers of Mathematics Principles and Standards for School Mathematics in the USA, states:

> Developing fluency requires a balance and connection between conceptual understanding and computational proficiency. On the one hand, computational methods that are over-practiced without understanding are often forgotten or remembered incorrectly. On the other hand, understanding without fluency can inhibit the problem solving process. (Smith, 2014)

Proficiency in algebra may be increased when instructional practices support the development of not only conceptual understanding in algebra but also procedural fluency (Kilpatrick, Swafford & Findell, 2001). It is to be hoped that teaching that

Fig. 8.9 First and second order generalisation

10^3 10^2 10^1 10^0 10^{-1} 10^{-2}	10^m
6^3 6^2 6^1 6^0 6^{-1} 6^{-2}	6^m
7^3 7^2 7^1 7^0 7^{-1} 7^{-2}	7^m
5^3 25^2 25^1 25^0 25^{-1} 25^{-2}	25^m
	p^m

10^3 6^3 7^3 5^3	10^2 6^2 7^2 5^2	10^1 6^1 7^1 5^1	10^0 6^0 7^0 5^0	10^{-1} 10^{-1} 10^{-1} 10^{-1}	
p^3	p^2	p^1	p^0	$p-1$	p^m

provides opportunities for students to make sense of algebraic symbols and procedures will also promote procedural fluency and conceptual understanding (Smith, 2014). As we have seen this has been demonstrated on a small scale in the results mentioned above.

Learning 'just algebra' is crucial to students' transition to tertiary courses, especially the STEM options. The two key ideas presented in this chapter are the various uses of letters and the concept of exponentiation. We believe it is an understatement to say that teachers need to be aware of the various meanings that students bring to algebra. One aim of education should be to prepare students for long-term understanding and success in algebra, and so much of the groundwork needs to be established and strengthened in the middle and lower secondary years of schooling. Hence, teachers have a responsibility to make sure that students' initial experiences of exponentiation and use of letters in algebra (including supportive met-befores) establish the basis for as clear an algebraic understanding as possible. It seems that combining historical ideas and current didactical constructs may reveal fresh approaches to the understanding of, and notation for, variables and powers in algebra.

<table>
<tr><td>1. Reading and writing powers of ten and their names in various ways. E.g. trillion= 10^{12} or 1000 000 000 000 or 10×10×10.....</td><td rowspan="9">Understanding Large Numbers Including Powers</td></tr>
<tr><td>2. Reading and writing large numbers. E.g. 532 609 418 056 or 532 billion 609 million.................. or $5\times10^{11}+3\times10^{10}+2\times10^{9}$.................</td></tr>
<tr><td>3. Grouping experiences (with matchsticks or craftsticks) and exponential notation: 4+4+4+4+4=4×5 , $4\times4\times4\times4\times4=4^5$, $10\times10\times10=10^3$</td></tr>
<tr><td>4. Working with large numbers and increasing quantity sense: How long does it take to count to a million, and to a billion, etc? How many grains of sand in all the beaches of the world?</td></tr>
<tr><td>5. Generating problems with a potential for calculation: If on average, a Year 9 student speaks 15000 words per day, how many words will be spoken by 210 students in 3 years?, How high is a million dollars in 1000 dollar bills? A billion dollars?</td></tr>
<tr><td>6. Calculator work: Comparing values such as 4×6 and 4^6, 7×6 and 7^6, and, 4^6 and 6^4, 3^5 and 5^3 . Also, evaluating 4^7, 4^9, 4^{16}, 4^{22}, 4^{29}.... and 4^6, 5^6, 9^6, 15^6 etc.</td></tr>
<tr><td>7. Authentic problems across content(measurement), and cross-curricular authentic word problems (e.g. science, social studies) involving large (& small) numbers and powers</td></tr>
<tr><td>8. Generalisation of exponential numerals to p^m</td></tr>
<tr><td>9. Opportunities to explore non-positive, and rational exponents such as the plant growth problem</td></tr>
</table>

Fig. 8.10 A framework for teaching the exponentiation concept

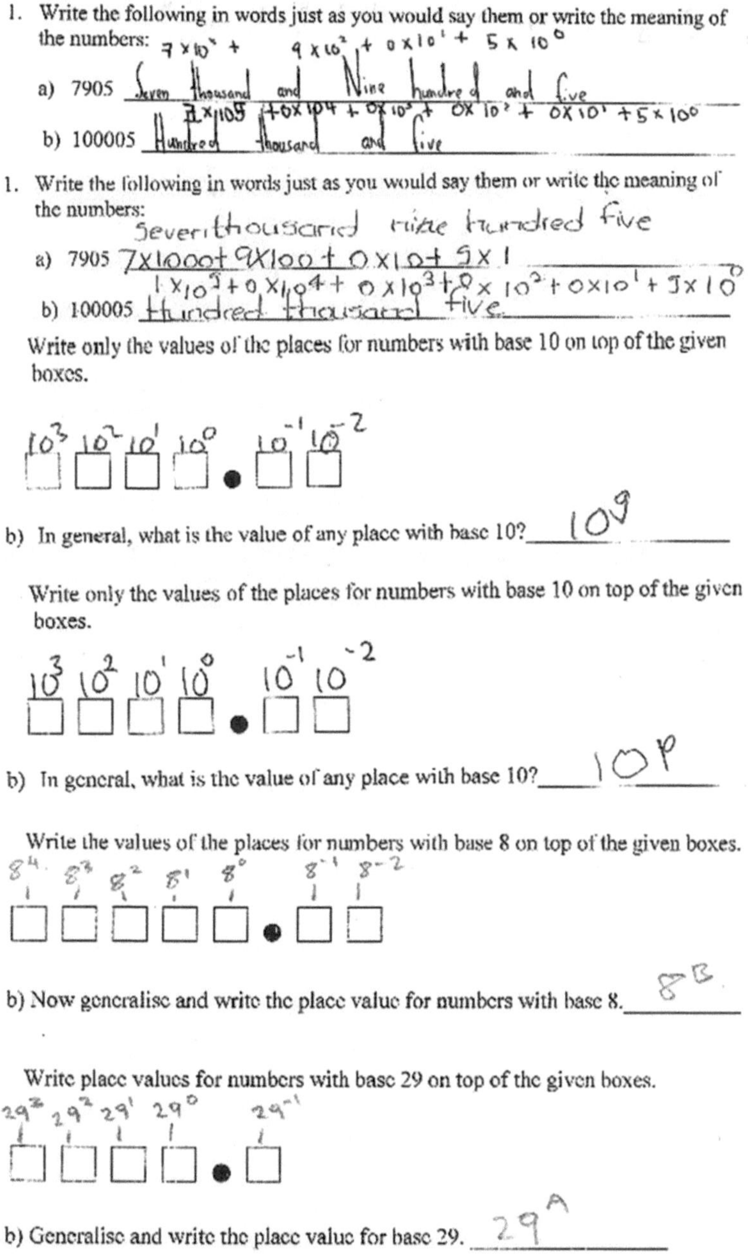

1. Write the following in words just as you would say them or write the meaning of the numbers: 7 x 10³ + 9 x 10² + 0 x 10¹ + 5 x 10⁰

a) 7905 Seven thousand and Nine hundred and five

1 x 10⁵ + 0 x 10⁴ + 0 x 10³ + 0 x 10² + 0 x 10¹ + 5 x 10⁰

b) 100005 Hundred thousand and five

1. Write the following in words just as you would say them or write the meaning of the numbers:

Seven thousand nine hundred five

a) 7905 7x1000 + 9x100 + 0x10 + 5x1

1 x 10⁵ + 0 x 10⁴ + 0 x 10³ + 0 x 10² + 0 x 10¹ + 5 x 10⁰

b) 100005 Hundred thousand five

Write only the values of the places for numbers with base 10 on top of the given boxes.

10³ 10² 10¹ 10⁰ . 10⁻¹ 10⁻²

b) In general, what is the value of any place with base 10? 10ᵍ

Write only the values of the places for numbers with base 10 on top of the given boxes.

10³ 10² 10¹ 10⁰ . 10⁻¹ 10⁻²

b) In general, what is the value of any place with base 10? 10ᴾ

Write the values of the places for numbers with base 8 on top of the given boxes.

8⁴ 8³ 8² 8¹ 8⁰ . 8⁻¹ 8⁻²

b) Now generalise and write the place value for numbers with base 8. 8ᴮ

Write place values for numbers with base 29 on top of the given boxes.

29³ 29² 29¹ 29⁰ . 29⁻¹

b) Generalise and write the place value for base 29. 29ᴬ

Fig. 8.11 Student answers showing use of exponentiation

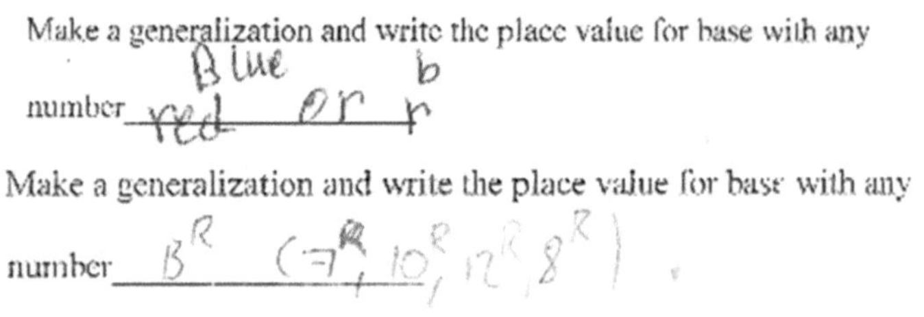

Fig. 8.12 Student answers showing generalisation in exponentiation

References

Arnon, A., Cottrill, J., Dubinsky, E., Oktac, A., Roa Fuentes, S., Trigueros, M., et al. (2014). *APOS theory: A framework for research and curriculum development in mathematics education*. New York: Springer.

Bag, A. K., & Sarma, S. R. (Eds.). (2003). *The concept of sunya*. New Delhi, India: IGNCA, INSA and Aryan Books International.

Bednarz, N., Kieran, C., & Lee, L. (Eds.). (1996). *Approaches to algebra: Perspectives for research and teaching*. Dordrecht, The Netherlands: Kluwer Academic.

Cajori, F. (1919). *The history of mathematics*. New York: The Macmillan Company.

Cangelosi, R., Madrid, S., Cooper, S., Olson, J., & Hartter, B. (2013). The negative sign and exponential expressions: Unveiling students' persistent errors and misconceptions. *The Journal of Mathematical Behavior, 32*(1), 69–82.

Colebrooke, H. T. (1817). *Classics of Indian mathematics: Algebra, with arithmetic and mensuration, from the Sanskrit of Brahmagupta and Bhaskara*. New Delhi, India: Sharada Publishing House.

Datta, B., & Singh, A. N. (2001). *History of Hindu mathematics* (Vols. 1–2). Bombay, India: Asia Publishing House.

Dubinsky, E., & McDonald, M. (2001). APOS: A constructivist theory of learning in undergraduate mathematics education research. In D. Holton, M. Artigue, U. Krichgraber, J. Hillel, M. Niss, & A. Schoenfeld (Eds.), *The teaching and learning of mathematics at university level: An ICMI study* (pp. 273–280). Dordrecht: Kluwer.

Eves, H. (1969). *An introduction to the history of mathematics*. Stillwater, ME: Holt, Rinehart and Winston.

Fauvel, J., & van Maanen, J. (Eds.). (2000). *History in mathematics education: The ICMI study*. Dordrecht, The Netherlands: Kluwer.

Gallardo, A. (2001). Historical-epistemological analysis in mathematics education: Two works in the didactics of algebra. In R. Sutherland, T. Rojano, A. Bell, & R. Lins (Eds.), *Perspectives on school algebra* (pp. 121–140). Dordrecht, The Netherlands: Kluwer.

Gallardo, A. (2008). Historical epistemological analysis in mathematics education: Negative numbers and the nothingness. In O. Figueras, J. L. Cortina, S. Alatorre, T. Rojano, & A. Sepulveda (Eds.), *Proceedings of the joint meeting of PME 32 and PME-NA XXX* (Vol. 1, pp. 17–29). Morelia, Mexico: Cinvestav-UMSNH.

Graham, A., & Thomas, M. O. J. (2000). Building a versatile understanding of algebraic variables with a graphic calculator. *Educational Studies in Mathematics, 41*, 265–282.

Gupta, R. C. (1987). One, two, three to infinity. *The Mathematics Teacher (India), 23*(4), 5–12.

Harper, E. (1987). Ghosts of Diophantus. *Educational Studies in Mathematics, 18*, 75–90.

Hiebert, J., & Lefevre, P. (1986). Procedural and conceptual knowledge. In J. Hiebert (Ed.), *Conceptual and procedural knowledge: The case of mathematics* (pp. 1–27). Hillsdale, NJ: Erlbaum.

Jhagroo, J., & Nataraj, M. (2015). When is a number too big to know? Scaffolding understanding of large number concepts. In R. Averill (Ed.), *Mathematics and statistics in the middle years: Evidence and practice* (pp. 162–181). Wellington, New Zealand: New Zealand Council for Educational Research.

Joseph, G. G. (2011). *The crest of the peacock: Non-European roots of mathematics*. Princeton, NJ: Princeton University Press.

Katz, V. (2007). Stages in the history of algebra with implications for teaching. *Educational Studies in Mathematics, 66*, 185–201.

Kieran, C. (1992). The learning and teaching of school algebra. In D. A. Grouws (Ed.), *Handbook of research on mathematics teaching and learning* (pp. 390–419). New York: Macmillan.

Kilpatrick, J., Swafford, J., & Findell, B. (Eds.). (2001). *Adding it up: Helping children learn mathematics*. Washington, DC: National Academy Press.

Knuth, E., Alibali, M., McNeil, N., Weinberg, A., & Stephens, A. (2005). Middle school students' understanding of core algebraic concepts: Equivalence & variable. *ZDM, 37*(1), 68–76.

Küchemann, D. E. (1981). Algebra. In K. M. Hart (Ed.), *Children's understanding of mathematics: 11-16* (pp. 102–119). London: John Murray.

Lee, L., & Wheeler, D. (1989). The arithmetic connection. *Educational Studies in Mathematics, 20* (1), 41–54.

MacGregor, M., & Stacey, K. (1997). Students' understanding of algebraic notation: 11-15. *Educational Studies in Mathematics, 33*(1), 1–19.

Malisani, E., & Spagnolo, F. (2009). From arithmetical thought to algebraic thought: The role of the "variable". *Educational Studies in Mathematics, 71*, 19–41.

Mason, J. (1996). Expressing generality and roots of algebra. In N. Bednarz, C. Kieran, & L. Lee (Eds.), *Approaches to algebra: Perspectives for research and teaching* (pp. 65–86). Dordrecht, The Netherlands: Kluwer.

Mason, J. (2004). *Doing ≠ construing and doing + discussing ≠ learning: The importance of the structure of attention*. Paper presented at the ICME10 Conference, Copenhagen, Denmark. Retrieved May 20, 2010, from http://math.unipa.it/~grim/YESS-5/ICME%2010%20Lecture%20Expanded.pdf

Mason, J., Graham, A., & Johnston-Wilder, S. (2005). *Developing thinking in algebra*. London: Sage.

Menninger, K. (1969). *Number words and number symbols* (P. Broneer, Trans.). Cambridge, MA: MIT Press. (Original work published 1958)

Nataraj, M. (2012). *Incorporating ideas from Indian history in the teaching and learning of general place value system*. Unpublished PhD thesis, Auckland University.

NZQA. (2014). *Level 1 and Level 2 NCEA algebra examination papers*. Retrieved from http://www.nzqa.govt.nz/qualifications-standards/qualifications/ncea/subjects/mathematics/levels/

Piaget, J., & Garcia, R. (1989). *Psychogenesis and the history of science*. New York: Columbia University Press.

Pitta-Pantazi, D., Christou, C., & Zachariades, T. (2007). Secondary school students' levels of understanding in computing exponents. *Journal of Mathematical Behaviour, 26*, 301–311.

Plofker, K. (2009). *Mathematics in India*. Princeton, NJ: Princeton University Press.

Puig, L., & Rojano, T. (2004). The history of algebra in mathematics education. In K. Stacey, H. L. Chick, & M. Kendal (Eds.), *The future of the teaching and learning of algebra: The 12th ICMI study* (pp. 189–223). Dordrecht, The Netherlands: Kluwer.

Radford, L. (1995). Before the other unknowns were invented: Didactic inquiries on the methods and problems of mediaeval Italian algebra. *For the Learning of Mathematics, 15*(3), 28–38.

Radford, L. (1996). Some reflections on teaching algebra through generalisation. In N. Bednarz, C. Kieran, & L. Lee (Eds.), *Approaches to algebra: Perspectives for research and teaching* (pp. 107–111). Dordrecht, The Netherlands: Kluwer.

Radford, L. (2000). Historical formation and student understanding of mathematics. In J. Fauvel & J. Van Maanen (Eds.), *History in mathematics education: The ICMI study* (pp. 143–170). Dordrecht, The Netherlands: Kluwer.

Schmittau, J. (1993). Retrieved from http://www.mlrg.org/proc3pdfs/Schmittau_Mathematics.pdf
Schoenfeld, A. H., & Arcavi, A. (1988). On the meaning of variable. *Mathematics Teacher, 81*(6), 420–427.
Sfard, A. (1995). The development of algebra: Confronting historical and psychological perspectives. *Journal of Mathematical Behavior, 14*, 15–19.
Shell Centre. (2015). Retrieved from http://map.mathshell.org/download.php?fileid=1668
Skemp, R. R. (1976). Relational understanding and instrumental understanding. *Mathematics Teacher, 77*, 20–26.
Smith T. M. (2014). *Instructional practices to support student success in Algebra 1*. Retrieved from https://www2.ed.gov/programs/dropout/instructionalpractices092414.pdf
Srinivasan, P. K. (1989). *Algebra for primary school: From class three: Three months correspondence course*. Unpublished paper, Madras, India.
Srinivasiengar, C. N. (1967). *The history of ancient Indian mathematics*. Calcutta, India: World Press Private.
Subramaniam, K., & Banerjee, R. (2011). The arithmetic-algebra connection: A historical pedagogical perspective. In J. Cai & E. Knuth (Eds.), *Early algebraization: A global dialogue from multiple perspectives* (pp. 85–107). New York: Springer.
Tall, D. O. (2004). The three worlds of mathematics. *For the Learning of Mathematics, 23*(3), 29–33.
Ursini, S. (2001). General methods: A way of entering the world of algebra. In R. Sutherland, T. Rojano, A. Bell, & R. Lins (Eds.), *Perspectives on school algebra* (pp. 209–230). Dordrecht, The Netherlands: Kluwer.
Ursini, S., & Trigueros, M. (2004). How do high school students interpret parameters in algebra? In M. J. Hoines & A. B. Fuglestad (Eds.), *Proceedings of the 28th Conference of the International Group for the Psychology of Mathematics Education* (Vol. 4, pp. 361–368). Bergen, Norway: Psychology of Mathematics Education.
Usiskin, Z. (1988). Conceptions of school algebra and uses of variables. In A. F. Coxford & A. P. Shulte (Eds.), *The ideas of algebra, K-12* (pp. 7–13). Reston, VA: National Council of Teachers of Mathematics.
van Amerom, B. (2002). *Reinvention of early algebra: Developmental research on the transition from arithmetic to algebra*. Unpublished doctoral thesis, Centre for Science and Mathematics Education, Utrecht University, The Netherlands.
Van Der Waerden, B. L. (1954). *Science awakening*. Groningen, The Netherlands: P. Noordhoff.
Vygotsky, L. S. (1978). *Mind in society: The development of higher psychological processes*. Cambridge, MA: Harvard University Press.
Wagner, D., & Davis, B. (2010). Feeling number: Grounding number sense in a sense of quantity. *Educational Studies in Mathematics, 74*, 39–51.
Weber, K. (2002a). Developing students' understanding of exponents and logarithms. In *Proceedings of the 24 Annual Meeting of the North American Chapter of Mathematics Education*. Retrieved from http://eric.ed.gov.ERICDocs/data/ericdocs2/content_storage_01/0000000b/80/27/e8/b5.pdf
Weber, K. (2002b). Students' understanding of exponential and logarithmic functions. In *Proceedings from the 2nd International Conference on the Teaching of Mathematics*. Retrieved from http://www.eric.ed.gov/PDFS/ED477690.pdf

Part IV
Proposed Future Developments

Chapter 9
Cognitive Neuroscience and Algebra: Challenging Some Traditional Beliefs

Carolyn Kieran

Abstract Recent studies using neuroimaging technology with tasks touching on various areas of mathematics are raising a great deal of excitement with their findings. This chapter presents some key work related to higher level mathematical reasoning and a few insights arising from these studies with respect to our current understanding of algebra learning. After a general introduction on cognitive neuroscience and its recent advances relevant to mathematics education, the chapter focuses on two studies in particular, one on the algebraic solving method and the other on representing functions. The chapter concludes with a discussion of the ways in which these results from the newly emerging field, which is at times referred to as mathematics educational neuroscience, offer the potential of casting a quite different light on how we think about students' processing of algebra-related material.

Keywords Cognitive neuroscience • Algebra • Functions • Symbolic method • Model method • Excelling in algebra

9.1 Introductory Remarks

Research on the development of algebraic skills, and their underlying conceptual foundations, has been an area of international interest since the late 1970s (see, e.g., Kieran, 1992, 2007). The body of research findings that has resulted from this interest has provided valuable empirical information on the various processes engaged in by students in their learning of algebra. It has also yielded theoretical constructs for interpreting these processes, as well as insights into the role that technological tools, various teaching approaches, and specific tasks can play in the development of that learning. Despite these advances in the body of knowledge related to algebra learning and teaching, cognitive neuroscience and neuroimaging data provide new tools for an even better understanding of the processing of mathematical tasks. While some of the findings of recent cognitive neuroscience

C. Kieran (✉)
Département de Mathématiques, Université du Québec à Montréal, Montreal, QC, Canada
e-mail: kieran.carolyn@uqam.ca

S. Stewart (ed.), *And the Rest is Just Algebra*, DOI 10.1007/978-3-319-45053-7_9

research are corroborating what we think we already know about mathematical learning, other findings are proving to be much more of a surprise to mathematics teachers and mathematics education researchers. It is the surprising findings—findings related to the processing of algebra- and function-related material by young adults—that are the main focus of this chapter. The chapter begins with a nontechnical introduction to the field of cognitive neuroscience and presents some of its recent results related to the processing of mathematical tasks. The section that follows goes into the details and findings of two cognitive neuroscience studies that are the core of this chapter. The chapter concludes with a discussion of these findings and relates them to the body of existing research evidence on algebra learning.

9.2 Cognitive Neuroscience and Mathematical Reasoning

9.2.1 Cognitive Neuroscience

Cognitive neuroscience, according to Wikipedia, is an academic field concerned with the scientific study of the biological substrates underlying cognition, with a specific focus on the neural substrates of mental processes, and addresses the questions of how psychological/cognitive functions are produced by neural circuits of the brain. Cognitive neuroscience is a branch of both psychology and neuroscience and relies upon theories drawn principally from cognitive science, but also overlaps with disciplines such as physiological psychology, cognitive psychology, and neuropsychology.

One of the main advances in this area of study occurred during the second half of the nineteenth century with the emergence of localization theories of the brain: the notion that different mental functions were related to specific areas of the brain. The twentieth century brought the phenomena of memory and thought and the cognitive revolution to the field. The demonstration that behavioral data do not provide enough information by themselves to explain mental processes led to the investigation of neural bases of behavior. Concurrent with the cognitive science movement, which was born in 1956 at a meeting at MIT where Chomsky and the team of Newell and Simon presented their work, neuroscience was established as a unified discipline in 1971. Interactions between neuroscience and cognitive science began to occur at the end of the 1970s and the term cognitive neuroscience was coined. The newly developed theories of cognitive science were adopted by cognitive neuroscience. As brain mapping technologies such as fMRI evolved, researchers began to use these technologies and the strategies of cognitive psychology to study brain function. The new field of cognitive neuroscience brought mind and brain together.

Quite recently, education was added to the mix to yield “mind, brain, and education” (Fischer, 2009). Educational neuroscience presently gathers together researchers in cognitive neuroscience, educational psychology, educational technology, and other related disciplines to explore the interactions between biological processes and education. A major goal of educational neuroscience is to bridge the gap between the two fields, with each field contributing to the other. Special Interest Groups (SIGs) devoted to neuroscience and education have been set up within educational research associations such as EARLI (SIG 22 met for the first time in Zürich in 2010), AERA, and BERA. While there has existed a certain amount of controversy as to whether cognitive neuroscience has a role to play in the broader field of education (e.g., Bruer, 1997), cognitive neuroscience has already made discoveries of use to education in general and to mathematics education in particular—mathematics educational neuroscience being considered a branch of educational neuroscience (see Campbell, 2010).

The two studies that are summarized in this chapter—studies where education researchers equipped with expert knowledge of the learning of mathematics have collaborated with cognitive science researchers—are but a few that illustrate the potential of mathematics educational neuroscience. The year 2010 also marked the first time that a leading mathematics education journal (*ZDM: The International Journal on Mathematics Education*) published a special issue containing a collection of studies that used neuroscientific methods to examine mathematics learning across a range of school levels (see, e.g., De Smedt & Verschaffel, 2010). A further indication of the emergence of this field is a chapter titled Mathematics Educational Neuroscience, which will appear in the upcoming *Third Handbook of Research on Mathematics Teaching and Learning*.

9.2.2 The Methods Used in Cognitive Neuroscience Research

Cognitive neuroscience research seeks to identify the brain activations that accompany elementary psychological processes. Among brain imaging methods, the various alternatives include functional Magnetic Resonance Imaging (fMRI), event-related potentials (ERP), electroencephalography (EEG), and near-infrared spectroscopy (NIRS). The fMRI method (see, e.g., Hernandez-García, Wager, & Jonides, 2002) uses MRI technology to measure brain activity by detecting changes in blood flow. The underlying principle is that cerebral blood flow and neural activity are related: when an area of the brain is activated, blood flow to that region increases. Increases in the amount of oxygenated blood are reflected in the magnetic properties of the blood. While fMRI typically has very good spatial resolution, it is relatively poorer with respect to temporal resolution because of the time required for blood flow to reach its peak in response to a given task. In contrast, the ERP method obtains reliable temporal readings of the physiological correlates of cognitive activity by means of EEG, which measures electrical activity of the brain over time using electrodes placed on the scalp. This technique allows accurate timing of changes in brain activity during execution of a cognitive task. While ERPs provide excellent temporal resolution, the dimension of spatial resolution is undefined.

Nevertheless, significant changes in the electrical activity recorded at each of the multiple electrode sites as subjects engage in a given task can yield general indications about the location of the neural structures being activated. The NIRS method, which has only recently begun to be used in educational settings (e.g., Obersteiner et al., 2010), involves placing a probe set on a participant's head and using near-infrared light to continuously measure changes in cerebral hemoglobin concentration. NIRS has lower spatial resolution than FMRI and lower temporal resolution than EEG, but higher temporal resolution than fMRI. While its portability and less restrictive nature make NIRS a practical option for school settings, this optical imaging technique is restricted to measuring cortical activity and not the subcortical activity that can be detected by fMRI. In addition to the various brain imaging technologies currently being used in cognitive neuroscience, eye-tracking technologies are also receiving increasingly widespread attention (e.g., Susac, Bubic, Kaponja, Planinic, & Palmovic, 2014).

Decisions regarding the appropriate imaging technology to be used in a given study must be aligned with the specific hypotheses to be tested, the choice of tasks to be used, the nature of the inferences to be drawn, and the suitability of various experimental designs, including the techniques of statistical analysis that will be applied to the data. While these various aspects are not the focus of this chapter, it is however noted that the design of cognitive neuroscience studies almost always provides for obtaining both neural and behavioral data. As emphasized by De Smedt et al. (2011), "the collection and analysis of behavioral data represents a necessary step in most fMRI experiments . . . and studies in cognitive neuroscience are grounded in hypotheses that are derived from behavioral (cognitive) data; in cognitive neuroscience, behavioral and neuroimaging data are considered on a level playing field with each type of data providing information that constrains the insights gleaned from the other, thereby becoming inextricably linked . . . an appreciation of multiple sources of data at different levels of description is essential to better understand a phenomenon under investigation" (p. 234). Often the behavioral data of cognitive neuroscience studies comprise accuracy rates and reaction times. While the articulation of the behavioral and neural data is central to the analyses of cognitive neuroscience studies, it is the neural data that provide information that is simply not discoverable by means of the behavioral data alone.

9.2.3 Cognitive Neuroscience and Arithmetic

The bulk of the recent work in cognitive neuroscience that has focused on mathematical reasoning has been related to the processes involved in arithmetic problem solving and reasoning. According to a review by Menon (2010), this research, which has examined various aspects of arithmetic processing such as retrieval, computation, and reasoning and decision making about arithmetic relations, has "helped to clarify which brain areas are critically and consistently engaged during arithmetic tasks, which regions provide a supportive role in arithmetic, and which brain areas contribute to arithmetic learning" (p. 515). For example, the research of Dehaene, Piazza, Pinel, and Cohen (2003) emphasized the role of the parietal cortex in number processing and arithmetic calculations. The parietal cortex has been found to be involved too in more complex mathematical processing such as word problem solving (Newman, Willoughby, & Pruce, 2011). Much of this cognitive neuroscience research has built upon prior psychological research that has identified some of the cognitive processes involved in learning arithmetic.

An example of a study that has examined the neural bases of psychological research findings in arithmetic is one that has been reported by Dresler et al. (2009) and Obersteiner et al. (2010). The study was designed to investigate whether pupil age and problem format (numeric or word format) would lead to different neural processing. The researchers, who were also interested in testing the feasibility of using NIRS technology with school children, conducted a school-based study involving 90 pupils from the 4th and 8th grades. Individual participants sat in front of a computer screen with probe sets placed on their heads. Error rates and reaction times served as behavioral data; for the NIRS data, the researchers focused on the oxygenated blood levels in certain regions of interest. The behavioral data showed that error rates were low and that, although numerical tasks were solved or read more quickly than word problems for both age groups, the difference between word problems and numerical problems was much smaller for grade 8 than for grade 4 students. The NIRS data revealed, as expected, that calculation resulted in slightly greater average oxygenation than did reading in parietal and posterior frontal regions. Surprisingly, similar brain activation patterns were found for both age groups. In view of the fact that older children had been hypothesized to use retrieval strategies to a greater amount and that they were more advanced in solving word problems due to their greater experience, the researchers expected to see more activation of brain areas associated with retrieval for the older students; but this was not the case. Their results led to the conclusion that the similarity of the activation patterns among 4th and 8th graders suggests that "while complex mental arithmetic [involving two-digit addition in the arithmetic problems] may develop from primary to secondary school in terms of more speeded calculation, these processes seem to rely on the parietal cortex in both age groups" (Obersteiner et al., 2010, p. 548). This study, which was the first to assess such a large number of students in a short time (14 days) within an educational setting, also illustrated the potential of the NIRS technology for studying young children's arithmetical activity.

9.3 Two Recent Cognitive Neuroscience Studies on the Processing of Algebra-Related Material

Up until the last decade, the processing of algebra-related material was extremely rare in cognitive neuroscience research. The two studies that are highlighted in this section are not the only ones to have been conducted on algebra topics, but they both serve to underline a central aspect related to the learning of algebra. Algebra is cognitively demanding! The studies provide clear evidence of the cognitive effort involved in doing and in being successful at algebra.

9.3.1 *Greater Cognitive Processing Required for Symbolic Method Than for Model Method*

In Singapore, students are taught in primary school how to solve word problems by means of the model method. Let us take, for example, the Sharing Problem illustrated in Fig. 9.1 (Ng, 2004), along with its solution by means of the model method.

A central feature of the model method is that unknowns are represented by unit rectangles. In this example, Samy's rectangle or unit is the generator of all the relationships presented in the problem. Raju, who has $100 more than Samy, is represented by a unit that is identical to Samy's, plus another different-sized rectangle representing the relational portion of $100 more. A model representation is formed that involves the two basic units, the $100 rectangle, and the total amount of $410. The entire structure of the drawing is the model representation. Students generally process such a model by undoing operations that involve subtracting 100 from 410 and then dividing 310 by 2.

The corresponding literal-symbolic formulation, taught when students are in early secondary school, involves representing each of the units by the unknown x, within the algebraic equation $x+x+100=410$. Generating symbolic representations for word problems, and using syntactic methods for solving with these symbolic representations, then becomes the norm for working with algebra problems throughout the secondary school experience. However, some students seem to prefer to use hybrid forms involving both model and symbolic methods (Khng & Lee, 2009). The researchers whose cognitive neuroscience study is the focus of this section of the chapter (i.e., Lee et al., 2010) were interested in whether the model and symbolic methods draw on similar cognitive processes and impose similar cognitive demands.

Fig. 9.1 The sharing problem and its accompanying model method

Raju and Samy shared $410 between them. Raju received $100 more than Samy. How much money did Samy receive?
2 units = $410 - $100
= $310
1 unit = $155
Samy received $155

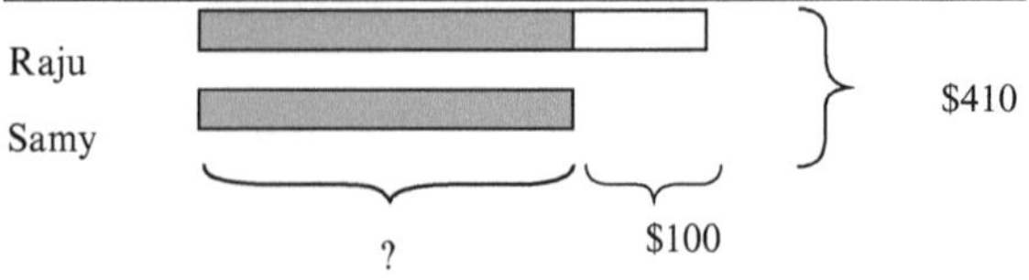

Lee et al. (2010) carried out their study with 17 right-handed adults (ten of whom were male), aged 22–29 years, who were proficient in both methods. The researchers used functional magnetic resonance imaging (fMRI) to examine whether the two methods involve similar cognitive processes and impose similar demands. Even if the solving time for the two methods were the same, fMRI would disclose whether different parts of the brain were activated by each method. In a previous study, the researchers (Lee et al., 2007) used fMRI to study the differences between the model and symbolic methods in the early stages of problem solving involving the transformation from text to either the model or the symbolic representation. They found that, while both methods were associated with activation of the working memory and quantitative processing regions of the brain, the symbolic method resulted in greater activity of those parts of the brain associated with attentional requirements. The 2010 study of Lee and his collaborators focused on the second stage of algebra word problem solving, that is, the computation of the actual solution to the problem from either the given model or the given symbolic representation.

While the advantages of fMRI are many, its constraints are such that the tasks need to be of short duration and typically involve key press responses. Sample tasks used in the Lee et al. (2010) study are shown in Fig. 9.2. It is noted that these tasks, while appearing to be quite simple in a school math context, were actually more complex than those typically encountered in fMRI studies. During trials involving the Model Experimental condition (ME), participants were presented with a model representation containing two rectangles, named J and M, with various relationships between the two being indicated. For the sample task shown in Fig. 9.2, J and M totaled 31 units, with M having 9 less than J. Participants were to find the number of units belonging to J. Trials involving the Symbolic Experimental condition (SE) were structurally identical to those given in the ME condition, but presented as algebraic equations. For the sample task shown in Fig. 9.2, J and M together added to 38, with M being 12 less than J. In order to control for processes related to the mere perception of model and symbolic representations, two control conditions (MC and SC) were used, with information presented in a manner that was not mathematically meaningful; participants were asked to take note of, and remember, the number that was in the same row as J (for the MC condition) or in the same row as "=J" or "J=" (for the SC condition).

The imaging technology used by the researchers allowed them to acquire both functional and structural brain images—the functional data yielding quite high temporal resolution and the structural data, high spatial resolution. As is the case with most studies of a cognitive neuroscience nature, Lee et al. (2010) present findings related to two types of data—those related to the study's behavioral data, for example, the correctness of the participants' responses, and those related to the imaging data. From the analysis of the behavioral data, it was found that participants were less accurate in the symbolic experimental condition (89 % success rate) than in the model experimental condition (96 % success rate), even though all participants had initially been screened to ensure that they could attain more than

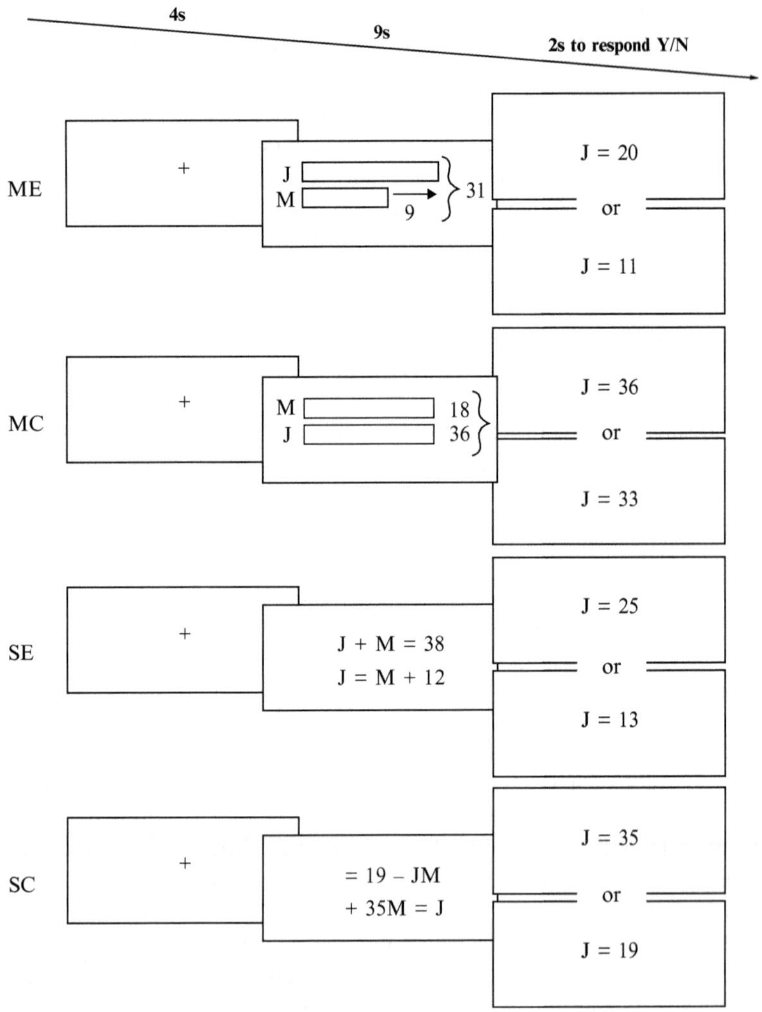

Fig. 9.2 Sequence of stimuli for the four conditions: model experimental (ME), model control (MC), symbolic experimental (SE), and symbolic control (SC). Participants first saw a fixation point for 4 s, followed by the problem, which was shown for 9 s. This was followed by the response screen. Participants were given 2 s to validate it against their own answers. Here, the response screen is illustrated with two alternatives (the *top slide* contains the correct answer). In the experiment, participants were provided with only one response alternative. Sample tasks from the Lee et al. (2010, p. 596) study (reprinted with permission from Springer)

90 % on problems similar to those used in the study and had less that 5 % difference in accuracy when using the two methods.

The imaging data were analyzed with respect to similarities and differences between the diagrammatic model and symbolic methods, as indicated by the areas of the brain that were activated by the calculation with each method. With respect to the symbolic method, greater activation was found in the middle and medial frontal

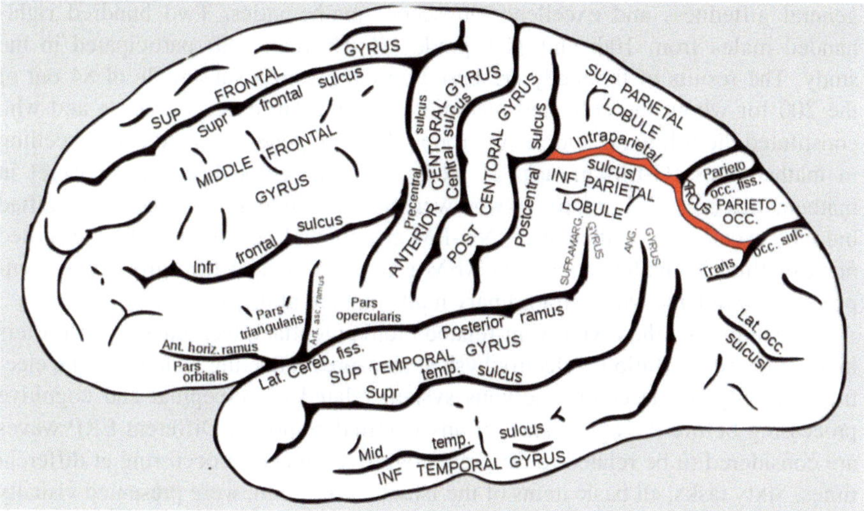

Fig. 9.3 Representation of lateral surface of left hemisphere of cerebral cortex (non-copyrighted material from the Internet)

gyri, anterior cingulate, caudate, precuneus, and intraparietal sulcus (see Fig. 9.3). The greater activation of these areas allowed the researchers to infer that additional attentional and executive resources are required for generating a numeric solution from an algebraic equation than from a diagrammatic model representation. The study also suggested that "linguistic processes play a more prominent role when processing symbolic stimuli" (Lee et al., 2010, p. 603). When the findings from this study were compared with those from their earlier study (which required only the translation of algebra word problems into such representations), as well as with related results from other recent cognitive neuroscience research, the results were consistent. The repeated finding that similar areas of the brain were differentially activated led the researchers to conclude that the symbolic method is more demanding than the diagrammatic model method. That the symbolic method is more effortful than the model method, even for competent adult algebra-problem-solvers, is a clear challenge to the traditional belief that algebraic methods of problem solving are easier than other methods and that algebraic solving activity is simply the mindless execution of an automatized set of techniques for symbol manipulation.

9.3.2 Cognitive Effort Required for Achieving Excellence in Algebra

A recent study by Waisman, Leikin, Shaul, and Leikin (2014) investigated the mathematical area of translation from graphical to symbolic representations of functions and their cerebral activation in groups of participants that differed in general giftedness and excellence in school mathematics. Two hundred right-handed males from 10th and 11th grade (16–18 years old) participated in the study. The results of the study are based on data from a subsample of 84 out of the 200 for whom the collected readings were without excessive noise and who constituted the following four main groups: 19 were generally gifted and excelling in mathematics (G-EM group), 21 were generally gifted but did not excel in mathematics (G-NEM group), 16 who were not identified as being generally gifted but who excelled in mathematics (NG-EM group), and 19 who were neither gifted nor excelling in mathematics (NG-NEM group). A special fifth group was composed of 9 students with extraordinary mathematical abilities (S-MG).

The study used the ERP (event-related brain potentials) technique, which offers high temporal resolution—electrophysical measures reflecting changes in the electrical activity of the central nervous system related to perceptual and cognitive processing before the appearance of any external response. Different ERP waves are considered to be related to different cognitive processes occurring at different times. Sixty tasks, all basic items of the Israeli curriculum, were presented visually to each participant. Each task was displayed in two consecutive windows with the graphical representation of the function followed by a suggested translation to symbolic form, to which the participant had to press a button on a keyboard as to whether the suggested symbolic representation was correct or not (see Fig. 9.4). Both behavioral analyses and electrophysiological analyses were carried out.

Behavioral analyses indicated that both G-EM and NG-EM students exhibited similar accuracy rates and reaction times. G-NEM students attained a level of accuracy similar to that of G-EM students by means of a longer reaction time devoted to the solving process. S-MG student were both significantly more accurate and quicker than students in both the G-EM and NG-EM groups.

Electrophysiological analyses indicated a greater latency at the early stage of perception for the S-MG group suggesting more complex mental activity early on for this group, followed by faster overall processing. That this group exhibited the highest accuracy along with the shortest reaction time is perhaps not so surprising. What is of more interest for this chapter is the finding that, among non-gifted students, those who excelled in mathematics achieved higher accuracy by means of greater mental effort. That is, NG-EM students had higher electrical brain activity than did the G-EM group across all time intervals. In fact, the mean amplitude levels for NG-EM students were the highest among the four participant groups (excluding the S-MG group).

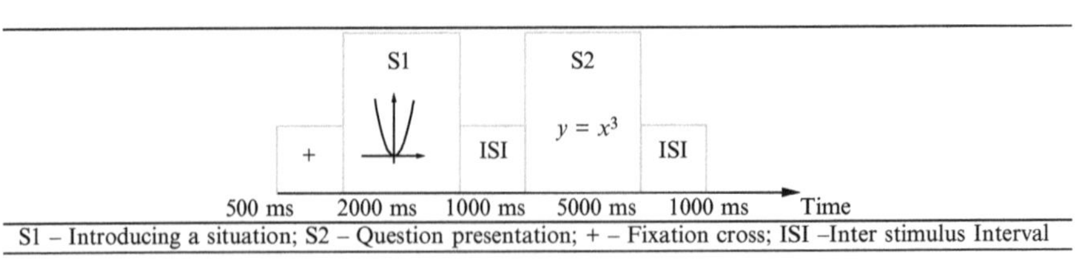

Fig. 9.4 The sequence of events and a task example from the Waisman et al. (2014, p. 676) study (reprinted with permission from Springer)

Waisman and her fellow researchers' finding that the students who were not generally gifted but who excelled in mathematics displayed the highest overall electrical activity of the four main groups is consistent with their results from a previous cognitive neuroscience study involving short insight-based problems (Leikin, Waisman, Shaul, & Leikin, 2012). As well, Waisman et al. (2014) found that only a combination of giftedness and mathematical excellence leads to lower cortical readings that, in turn, reflect lower cognitive load. In other words, prior expertise in problem solving does not necessarily lower the mental load. Thus, they argue that neurocognitive efficiency "does not characterize brain activity in all experts in problem solving and ... that problem-solving expertise developed by students without general giftedness is achieved by means of high cognitive effort" (Waisman et al., 2014, pp. 689–690). The researchers suggest that this effort may indicate that such students "allocate more mental resources for devoting their attention to the graph, to classify stimuli features and to retrieve relevant information (symbolic equation) from memory" (p. 690). However, the techniques and research design employed in the previously discussed study by Lee et al. (2010)—techniques that allowed them to confirm that additional attentional and executive resources are required for generating a numeric solution from an algebraic equation than from a diagrammatic model representation—suggest that it is not attention to the graph itself, but rather to its algebraic symbolic entanglements, that is at play here. The neural-data-based finding from the Waisman et al. study with respect to the cognitive effort expended by those who excel clearly challenges the naive belief that students who do well in algebra and functions are naturals at it and achieve that excellence without a great deal of cognitive effort.

9.4 Discussion

Despite the many caveats that could be raised with respect to the methods of cognitive neuroscience research and thus to the validity of the conclusions one might draw from such research (Turner, 2011), there is no question as to the insights into mathematical processing that this research provides, insights that could not otherwise be obtainable. Brain imaging techniques are able to yield information not discoverable by more traditional, behavioral, research methods. The limitations of existing behavioral and subjective self-report methods highlight the problem of using such methods to speculate about cognitive activity. As we have seen from the two main studies described in this chapter, one of the most compelling findings to emerge from this recent cognitive neuroscience research concerns the nature of algebraic processing. Not only does the algebraic method of equation formulation and equation solving require a great deal more cognitive attention than does the diagrammatic model method (Lee et al., 2007, 2010), but also that those who are not gifted mathematically but who excel in algebra achieve this excellence by means of a great deal of mental effort (Waisman et al., 2014).

These research findings that suggest that high cognitive effort is required in order to use algebraic methods and to excel at algebra—even for competent young adults—are so compelling because they go against the widespread view that algebraic manipulation is simply a mindless and highly automatized activity that involves executing algorithmic procedures for expression simplification and equation solving. Research on teachers' knowledge of students' algebraic thinking indicates that teachers (and researchers too) believe that the literal symbolic representation is much easier for students to handle than are other representations of algebra-related problems. For example, Nathan and Koedinger (2000) asked a group of 67 high school mathematics teachers and 35 mathematics education researchers to rank order 12 mathematics problems from easiest to most difficult. Four were in story-problem format (e.g., "When Ted got home from his waiter job, he multiplied his hourly wage by the six hours he worked that day. Then he added the $66 he made in tips and found he earned $81.90. How much per hour did Ted make?"), four were in symbol-equation format (e.g., "Solve for x: $x \cdot 6 + 66 = 81.90$"), and four were in word-equation format (e.g., "Starting with some number, if I multiply it by 6 and then add 66, I get 81.90. What did I start with?"). Teachers and researchers predicted that story problems and word-equation problems would be more difficult than symbol-equation problems.

When Koedinger and Nathan (2004) gave the same problems to a group of 76 high school students (Algebra I and post-Algebra I students), the symbol-equation format was found to be significantly less likely to be correctly solved than either the story-problem or word-equation formats. In a replication study involving 171 students (all Algebra I students) solution success rates for symbol-equation format were 25 % less than for story problems and nearly 20 % less than for word equations. The finding that teachers believed that symbol-equation problems would be the easiest for students and that, in fact, students found the symbol-equation format the most difficult is telling. Nathan and Petrosino (2003) have provided evidence to support the argument that it is the well-developed subject-matter knowledge of the high school teachers (i.e., the "expert blind spot" syndrome) that underpinned their view and that led them to inaccurately predict students' algebra problem-solving difficulty.

Koedinger and Nathan (2004) have emphasized that the learning of algebraic technique and the various subtleties involved in algebraic transformational activity takes much longer than teachers may realize. Other researchers have also noted the many aspects of algebra that students find difficult to master. For example, Hoch and Dreyfus (2004) observed that the 11th graders who participated in their study on the recognition of form within algebraic expressions and equations had a very poor sense of form, which led to inconsistent and erroneous symbol manipulation. Bloedy-Vinner (1994, 2001) investigated Israeli matriculation students' understanding of parameters and variables, as well as the notion that an algebraic letter that starts off as a parameter might change in meaning throughout the process of solving a problem. In understanding this difference in roles, implicit quantifiers are involved. From a questionnaire designed to test students' understanding of implicit quantifier structures, she found that most of the questions yielded very low success rates ranging from 3 to 69 %. Findings such as these have led some researchers (e.g., Artigue, 2002; Lagrange, 2000) to suggest that, for the older student of algebra, the use of technological tools, such as Computer Algebra Systems (CAS), can lead to the kind of conceptual development that is needed in order to successfully manipulate algebraic symbols—as long as the technical aspects are not ignored. In other words, with an eye to the development of algebraic technique, as well as an instructional practice that views the CAS not only as a utilitarian tool but also as a pedagogical tool, the growth of the conceptual and theoretical ideas that support algebraic technical facility can be fostered.

Past research, which has shown that students need a great deal of time in becoming comfortable with algebraic symbols and in acquiring the fluency and power that symbols can provide, has led to the suggestion that students ought to begin the process at an earlier age (Cai et al., 2005) and to studies that have yielded empirical evidence of the kinds of algebraic thinking that can be developed in primary school children (e.g., Blanton et al., 2015; Schliemann, Carraher, & Brizuela, 2012). While this emerging body of research illustrates children's ability to engage in algebraic thinking from as early as first grade (e.g., Blanton, Brizuela, Gardiner, Sawrey, & Newman-Owens, 2015), the cognitive neuroscience study of Lee et al. (2010) led the researchers to question whether algebraic activity is appropriate for the younger student. From their finding that the symbolic approach is more effortful even among competent adult problem solvers they concluded:

> In relation to the teaching of algebra, the model method is thought to provide children better access to algebra because it is less abstract and more visual than symbolic algebra. Our findings offer new insights into the reasons why many students find the model method easier. Contrary to expectations, we found no evidence that it relies more extensively on visual processes than does the symbolic method. Instead, we found that [the symbolic method] imposes greater demands on attentional resources. (p. 604)

De Smedt and Verschaffel (2010) drive home this point when they state that recent cognitive neuroscience research shows "that some solution methods are cognitively more demanding than others (Lee et al., 2010; Thomas, Wilson, Corballis, Lim, & Yoon, 2010); these data suggest that it might not be appropriate to teach these methods at young ages, when functions of working memory and attentional control have not fully developed yet (Luna, Garver, Urban, Lazar, & Sweeney, 2004)" (p. 651).

The two cognitive neuroscience studies that were highlighted in this chapter, those of Lee et al. (2010) and Waisman et al. (2014), serve to raise our awareness levels of certain cognitive constraints associated with the doing of algebra, even if the findings do not directly provide the tools and information needed to adapt teaching so as to take these constraints into consideration. More particularly, they serve to dispel the notion that algebraic manipulation can be treated as a mere postscript to the conceptual or technical lesson at hand. The finding that algebraic excellence requires a great deal of mindful attention and cognitive effort should obviously sensitize us teachers and researchers to the mental demands involved in doing algebra—and the research subjects of these studies were students who were very proficient in algebra. These findings should, at the very least, move us—we who are teachers of college level mathematics—to pause and reflect upon the phenomenon of students having difficulty with performing effectively on the algebraic components of the mathematical tasks we routinely put to them. The neurocognitive results discussed above should lead us to realize that, after we have presented some new higher level mathematics, which might include some "lower-level" algebraic activity, it is far from being the case that the algebraic part is "just algebra." We need to start questioning our traditional beliefs about students' algebraic activity—beliefs that include the notion that such activity is straightforward and requires nothing more than the application of well-learned algorithmic procedures. As Will Rogers, the American humorist and social commentator, once so famously remarked, it isn't what we don't know that gives us trouble, it's what we know that isn't so that gives us trouble.

References

Artigue, M. (2002). Learning mathematics in a CAS environment: The genesis of a reflection about instrumentation and the dialectics between technical and conceptual work. *International Journal of Computers for Mathematical Learning, 7*, 245–274.

Blanton, M., Brizuela, B. M., Gardiner, A. M., Sawrey, K., & Newman-Owens, A. (2015). A learning trajectory in six-year-olds' thinking about generalizing functional relationships. *Journal for Research in Mathematics Education, 46*, 511–558.

Blanton, M., Stephens, A., Knuth, E., Gardiner, A. M., Isler, I., & Kim, J.-S. (2015). The development of children's algebraic thinking: The impact of a comprehensive early algebra intervention in third grade. *Journal for Research in Mathematics Education, 46*, 39–87.

Bloedy-Vinner, H. (1994). The analgebraic mode of thinking: The case of parameter. In J. P. da Ponte & J. F. Matos (Eds.), *Proceedings of the 18th Conference of the International Group for the Psychology of Mathematics Education* (Vol. 2, pp. 88–95). Lisbon, Portugal: PME.

Bloedy-Vinner, H. (2001). Beyond unknowns and variables—Parameters and dummy variables in high school algebra. In R. Sutherland, T. Rojano, A. Bell, & R. Lins (Eds.), *Perspectives on school algebra* (pp. 177–189). Dordrecht, The Netherlands: Kluwer.

Bruer, J. T. (1997). Education and the brain: A bridge too far. *Educational Researcher, 26*(8), 4–16.

Cai, J., Lew, H. C., Morris, A., Moyer, J. C., Ng, S. F., & Schmittau, J. (2005). The development of students' algebraic thinking in earlier grades: A cross-cultural comparative perspective. *Zentralblatt für Didaktik der Mathematik, 37*, 5–15.

Campbell, S. R. (2010). Embodied minds and dancing brains: New opportunities for research in mathematics education. In B. Sriraman & L. English (Eds.), *Theories of mathematics education* (pp. 309–331). Berlin: Springer. doi:10.1007/978-3-642-00742-2_31.

De Smedt, B., Ansari, D., Grabner, R. H., Hannula-Sormunen, M., Schneider, M., & Verschaffel, L. (2011). Cognitive neuroscience meets mathematics education: It takes two to tango. *Educational Research Review, 6*, 232–237.

De Smedt, B., & Verschaffel, L. (2010). Traveling down the road: From cognitive neuroscience to mathematics education ... and back. *ZDM: The International Journal on Mathematics Education, 42*, 649–654. doi:10.1007/s11858-010-0282-5.

Dehaene, S., Piazza, M., Pinel, P., & Cohen, L. (2003). Three parietal circuits for number processing. *Cognitive Neuropsychology, 20*(3–6), 487–506.

Dresler, T., Obersteiner, A., Schecklmann, M., Vogel, A. C. M., Ehlis, A.-C., Richter, M. M., et al. (2009). Arithmetic tasks in different formats and their influence on behavior and brain oxygenation as assessed with near-infrared spectroscopy (NIRS): A study involving primary and secondary school children. *Journal of Neural Transmission, 12*(16), 1689–1700.

Fischer, K. W. (2009). Mind, brain, and education: Building a scientific groundwork for learning and teaching. *Mind, Brain and Education, 3*(1), 3–16. doi:10.1111/j.1751-228X.2008.01048.x.

Hernandez-García, L., Wager, T., & Jonides, J. (2002). Functional brain imaging. In H. Pashler & J. Wixted (Eds.), *Stevens' handbook of experimental psychology* (Methodology in experimental psychology 3rd ed., Vol. 4, pp. 175–221). New York: Wiley. http://onlinelibrary.wiley.com/doi/10.1002/0471214426.pas0405/full.

Hoch, M., & Dreyfus, T. (2004). Structure sense in high school algebra: The effects of brackets. In M. J. Hoines & A. B. Fuglestad (Eds.), *Proceedings of the 28th Conference of the International Group for the Psychology of Mathematics Education* (Vol. 3, pp. 49–56). Bergen, Norway: PME.

Khng, K. H., & Lee, K. (2009). Inhibiting interference from prior knowledge: Arithmetic intrusions in algebra word problem solving. *Learning and Individual Differences, 19*, 262–268.

Kieran, C. (1992). The learning and teaching of school algebra. In D. A. Grouws (Ed.), *Handbook of research on mathematics teaching and learning* (pp. 390–419). New York: Macmillan.

Kieran, C. (2007). Learning and teaching algebra at the middle school through college levels: Building meaning for symbols and their manipulation. In F. K. Lester Jr. (Ed.), *Second handbook of research on mathematics teaching and learning* (pp. 707–762). Charlotte, NC: Information Age.

Koedinger, K. R., & Nathan, M. J. (2004). The real story behind story problems: Effects of representations on quantitative reasoning. *Journal of the Learning Sciences, 13*, 129–164.

Lagrange, J.-B. (2000). L'intégration d'instruments informatiques dans l'enseignement : une approche par les techniques [The integration of computer tools into teaching: An approach according to techniques]. *Educational Studies in Mathematics, 43*, 1–30.

Lee, K., Lim, Z. Y., Yeong, S. H. M., Ng, S. F., Venkatraman, V., & Chee, M. W. L. (2007). Strategic differences in algebraic problem solving: Neuroanatomical correlates. *Brain Research, 1155*, 163–171.

Lee, K., Yeong, S. H. M., Ng, S. F., Venkatraman, V., Graham, S., & Chee, M. W. L. (2010). Computing solutions to algebraic problems using a symbolic versus a schematic strategy. *ZDM: The International Journal on Mathematics Education, 42*, 591–605. doi:10.1007/s11858-010-0265-6.

Leikin, R., Waisman, I., Shaul, S., & Leikin, M. (2012). An ERP study with gifted and excelling male adolescents: Solving short insight-based problems. In T. Y. Tso (Ed.), *Proceedings of the 36th International Conference for the Psychology of Mathematics Education* (Vol. 3, pp. 83–90). Taiwan, Taipei: PME.

Luna, B., Garver, K. E., Urban, T. A., Lazar, N. A., & Sweeney, J. A. (2004). Maturation of cognitive processes from late childhood to adulthood. *Child Development, 75*, 1357–1372.

Menon, V. (2010). Developmental cognitive neuroscience of arithmetic: Implications for learning and education. *ZDM Mathematics Education, 42*, 515–525. doi:10.1007/s11858-010-0242-0.

Nathan, M. J., & Koedinger, K. R. (2000). Teachers' and researchers' beliefs about the development of algebraic reasoning. *Journal for Research in Mathematics Education, 31*, 168–190.

Nathan, M. J., & Petrosino, A. (2003). Expert blind spot among preservice teachers. *American Educational Research Journal, 40*, 905–928.

Newman, S. D., Willoughby, G., & Pruce, B. (2011). The effect of problem structure on problem-solving: An fMRI study of word versus number problems. *Brain Research, 1410*, 77–88.

Ng, S. F. (2004). Developing algebraic thinking in early grades: Case study of the Singapore primary mathematics curriculum. *The Mathematics Educator, 8*(1), 39–59.

Obersteiner, A., Dresler, T., Reiss, K., Vogel, A. C. M., Pekrun, R., & Fallgatter, A. J. (2010). Bringing brain imaging to the school to assess arithmetic problem solving: Chances and limitations in combining educational and neuroscientific research. *ZDM—The International Journal on Mathematics Education, 42*, 541–554. doi:10.1007/s11858-010-0256-7.

Schliemann, A. D., Carraher, D. W., & Brizuela, B. M. (2012). Algebra in elementary school. *Enseignement de l'algèbre élémentaire* (Special Issue of *Recherches en Didactique des Mathé matiques)* (pp. 107–122).

Susac, A., Bubic, A., Kaponja, J., Planinic, M., & Palmovic, M. (2014). Eye movements reveal students' strategies in simple equation solving. *International Journal of Science and Mathematics Education, 12*, 555–577.

Thomas, M. J., Wilson, A. J., Corballis, M. C., Lim, V. K., & Yoon, C. (2010). Evidence from cognitive neuroscience for the role of graphical and algebraic representations in understanding function. *ZDM—The International Journal on Mathematics Education, 42*, 607–619. doi:10.1007/s11858-010-0272-7.

Turner, D. A. (2011). Which part of 'two way street' did you not understand? Redressing the balance of neuroscience and education. *Educational Research Review, 6*, 224–232.

Waisman, I., Leikin, M., Shaul, S., & Leikin, R. (2014). Brain activity associated with translation between graphical and symbolic representations of functions in generally gifted and excelling in mathematics adolescents. *International Journal of Science and Mathematics Education, 12*, 669–696.

Chapter 10
Rethinking Algebra: A Versatile Approach Integrating Digital Technology

Mike Thomas

Abstract Many have thought deeply about the construction of the school algebra curriculum, but the question remains as to why we teach the topics we do in the manner we do, stressing manipulations of symbols, and why some other avenues are ignored. In this chapter we consider the basic constructs in the school algebra curriculum and the procedural approach often taken to learning them and suggest some reasons why certain topics may be excluded. We examine how particular tasks, including some that integrate digital technology into student activity, could be used to rethink the algebra curriculum content with a view to motivating students and promoting versatile thinking. Some reasons why these topics have often not yet found their way into the curriculum are discussed.

Keywords Versatile thinking • Algebra • Tertiary • Digital technology • Representations

The aim of this chapter is to rethink both the content of secondary school algebra and the manner of its delivery and to ask: Should either, or both, be changed in order to improve understanding of algebra? There seems little doubt about two crucial statements:

- Algebra (including the school algebra of generalised arithmetic) is of fundamental importance in mathematics.
- Many students find most of school algebra either difficult or impossible to comprehend.

These two statements are linked together by the fact that school algebra is a semiotic system. It is the signs or representations of this system that at one and the same time make algebra so useful and yet so difficult for many. Consider, for example, the compressive power in a relatively simple symbolism

M. Thomas (✉)
Mathematics Department, The University of Auckland, Auckland, New Zealand
e-mail: moj.thomas@auckland.ac.nz

S. Stewart (ed.), *And the Rest is Just Algebra*, DOI 10.1007/978-3-319-45053-7_10

$$\sum_{i=0}^{i=2} w^i$$

where w is a cube root of unity. Suspending for a moment the fact that this summation comes to zero and ignoring simplifications of w^2, if we fully expand the symbolisation we get

$$\begin{aligned}\sum_{i=0}^{i=2} w^i &= 1 + w + w^2 = 1 + \frac{-1 + i\sqrt{3}}{2} + \left(\frac{-1 + i\sqrt{3}}{2}\right)^2 \\ &= 1 + \frac{-1 + \sqrt{-1}\sqrt{3}}{2} + \left(\frac{-1 + \sqrt{-1}\sqrt{3}}{2}\right)^2 \\ &= 1 + \frac{-1 + \sqrt{-1}\sqrt{3}}{2} + \left(\frac{-1 + \sqrt{-1}\sqrt{3}}{2}\right)\left(\frac{-1 + \sqrt{-1}\sqrt{3}}{2}\right)\end{aligned}$$

Mason (1987) agrees that a semiotic problem, concerning the relationship between the sign and the signified, or the symbol and the symbolised, is at the root of algebraic difficulties. This semiotic difficulty is not surprising when we consider how long it took for the symbolism to settle down into our modern version. For example, Struik (1969) gives these examples.

(a) What must be the amount of a square, which, when twenty-one dirhams are added to it becomes equal to the equivalent of ten roots of that square? Al-Khwarizmi ca. 825 AD
(b) cubus p : 6 rebus aequalis 20 Cardan ca. 1545 AD
(c) aaa - 3bba = +2ccc Harriot ca. 1610 AD

The triadic model of Peirce describes how signs, constructed through thoughts and ideas, comprise three components: the representamen [or the external material entity]; the object referred to; and the interpretant, or the sense made of the entity. Unlike icons and indexes, symbols, including those used in mathematics, have become associated with their meaning by accepted usage (Peirce, 1898). The grouping of these symbols into systems (sometimes called a representation system), such as the algebra of generalised arithmetic considered here, requires more than a set of symbols; it also needs rules for their production and transformation, and a set of relationships between the signs and their meanings (see Ernest, 2006). Student activity, both within such a system and converting between systems (Duval, 2006), can lead to key epistemological aspects and understanding, of mathematical objects, contributing to the goal of helping students attain *versatile thinking* in mathematics, which according to Thomas (2008a, 2008b), involves at least three abilities:

- To switch at will in any given representational system between a perception of a particular mathematical entity as a process and the perception of the entity as an object

- To exploit the power of visual schemas by linking them to relevant logico/analytic schemas
- To work seamlessly within and between representations, and to engage in procedural and conceptual interactions with representations

Thus a versatile view (Graham, Pfannkuch, & Thomas, 2009; Graham & Thomas, 2000, 2005; Tall & Thomas, 1991; Thomas, 1988, 2002, 2008a, 2008b) of the semiotic system of school algebra requires more than the ability to transform symbols according to the rules of the system; it also means making sense of them as processes and objects, and the ability to relate them to other systems. However, much of what happens in school algebra comprises activity aimed at transformations according to the rules of the system with much less effort addressed to considering sense making or conversions. Such standard manipulation algebra (Thomas & Tall, 2001) often leads to what Skemp (1976) described as instrumental understanding, or applying rules without clear reasons.

In order to be able to operate on an entity within a further process, such as when manipulating symbolic literals in algebra, APOS theory (Dubinsky, 1991) tells us that students need an object view of the symbols (although what kind of object they perceive is often open to question—see Tall, Thomas, Davis, Gray, & Simpson, 2000). While in the higher level mathematics of formal world thinking (Tall, 2004, 2008) objects can be brought into being through a definition, which specifies their properties, in school algebra students are often left to abstract properties of objects such as variable, expression, equation, function and polynomial for themselves by learning and repeating procedural actions on symbols. In this chapter I suggest that more attention could be paid to relating the algebraic symbols to other representations and investigating the properties of the objects of algebra. I also propose ways that this could be achieved by harnessing the investigative power of digital technology (DT).

10.1 A Theoretical Framework

In other papers we have proposed a Framework for Advanced Mathematical Thinking (FAMT) (Stewart & Thomas, 2010; Thomas & Stewart, 2011) that combines orthogonally elements of the action-process-object-schema (APOS) framework for studying learning, presented by Dubinsky and others (Dubinsky, 1991; Dubinsky & McDonald, 2001) with each of Tall's (2004, 2008) Three Worlds of Mathematical Thinking. APOS theory describes how mental objects may be constructed from actions and processes via reflective abstraction, while Tall's framework suggests that mathematical thinking can involve an embodied world, with its *visual* and *enactive* aspects, a symbolic world of semiotic symbols, and the formal world of axiomatic and deductive mathematics. The FAMT is based on the principle that each mathematical concept can be examined in terms of action, process and object types of thinking in each of the embodied and symbolic and

Worlds APOS	Embodied World	Symbolic World		Formal World
		Algebraic Rep.	Matrix Rep.	
Action	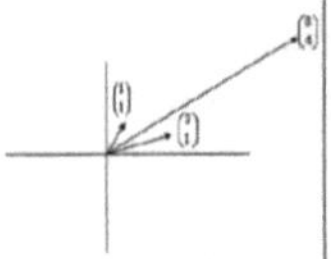 Can add multiples of two specific vectors	Can create a new vector **w** by, say addition. e.g **w** = 3**u** + 5**v**	Can calculate with linear combinations, e.g. $2\begin{bmatrix}2\\3\\1\end{bmatrix}+3\begin{bmatrix}4\\-1\\1\end{bmatrix}=\begin{bmatrix}16\\3\\5\end{bmatrix}$ Can determine whether a vector **w** is a linear combination of **u** and **v** using row reduction	
Process	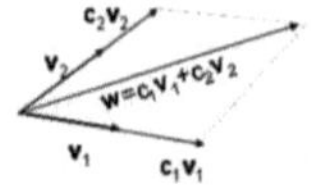 Can generalise addition of multiples of any vectors	Can think of linear combinations of vectors e.g. $\mathbf{w} = a\mathbf{u} + b\mathbf{v}$ without having to perform operations	Can consider operations on vectors without performing them e.g. $k_1\begin{pmatrix}x_1\\ \vdots\\ x_n\end{pmatrix}+k_2\begin{pmatrix}y_1\\ \vdots\\ y_n\end{pmatrix}$	Can relate linear combination to other linear algebra concepts such as span and linear independence
Object	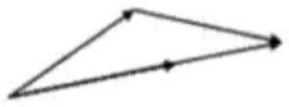 Sees resultant as new vector object and can operate on it	Can operate on a linear combination e.g. $T(a\mathbf{u} + b\mathbf{v})$	Can operate on a linear combination e.g. $A.\begin{pmatrix}k_1\mathbf{x}_1+k_2\mathbf{y}_1\\ \vdots\\ k_1\mathbf{x}_n+k_2\mathbf{y}_n\end{pmatrix}$	$\mathbf{w} = c_1\mathbf{v}_1 + c_2\mathbf{v}_2 + \ldots + c_k\mathbf{v}_k$ sees a general linear combination as an element of a vector space **V** $c_i \in F$

Fig. 10.1 The Framework for Advanced Mathematical Thinking (FAMT) applied to linear combination

formal worlds of mathematics. Hence, a matrix of cells may be produced with each cell targeting student thinking and understanding in one area, such as an embodied process. While we have found FAMT particularly useful for analysing student thinking in university mathematics, namely linear algebra (see the example in Fig. 10.1), the underlying principles may also prove useful in school level mathematics and we will consider this below.

Providing tasks that enable students to engage in activity that encourages them to think in the manner described by as many of the cells of the framework as possible for a given mathematical construct and to construct meaningful links between them, is one way to promote versatile thinking. This is a key tenet of the ideas presented here.

We will now look at some of the key ideas in school algebra and ask how DT might assist students to construct versatile thinking about them.

10.2 Variables and Expressions in School Algebra

The concept of variable is not an easy one for students to construct. Even Bertrand Russell found the notion of variable problematic.

> 6. Mathematical propositions are not only characterized by the fact that they assert implications, but also by the fact that they contain *variables*. The notion of the variable is one of the most difficult with which logic has to deal. For the present, I openly wish to make it plain that there are variables in all mathematical propositions, even where at first sight they might seem to be absent. . .We shall find always, in all mathematical propositions, that

the words *any* or *some* occur; and these words are the marks of a variable and a formal implication. (Russell, 1903, pp. 5, 6)

It has been known for well over 35 years now that students have problems understanding the use of symbolic literals or letters in algebra (Küchemann, 1981; Wagner, 1981). That these problems in understanding are persistent was shown by Küchemann's (1981) investigation into children's understanding of the use of letters in algebra, as part of the wide-ranging CSMS study, with four or five years of algebra teaching making very little difference to their understanding of the subject. Around 30 years later in a follow-up study (Hodgen, Brown, Küchemann, & Coe, 2010; Hodgen, Coe, Brown, & Küchemann, 2014) the group concluded that attainment had not changed very much, and

- Algebra results show fewer students reaching the higher Levels 3 and 4, which is the point at which students begin to understand the key algebraic concepts of variable and generalised number (Hodgen et al., 2010, p. 6)

Rosnick and Clement (1980) too showed that even college students had similar problems, such as confusing the use of letter as variable with the use as a label or unit. One of the factors causing this situation is the multiplicity of uses of letters in mathematics, with Wagner (1981) listing placeholder, index, specific unknown, generalised number, indeterminate, independent or dependent variable, constant and parameter as possible uses. She also pointed out that this complexity is increased by the fact that different letters can be used to represent the same thing, and the same letter can be used to represent different things. It still often seems to be the case that, as Skemp (1971, p. 227) noted, 'The idea of a variable is in fact a key concept in algebra—although many elementary texts do not explain or even mention it'. This omission of explaining what a variable is still extends to many classrooms. Hence, expecting students to abstract all the subtle complexities of symbolic literals simply from procedural use of letters appears to be a step too far.

The difficulties students experience with use of letters clearly impinges on the way they view symbols such as '$x + 3$'. Many will not accept this kind of expression as an answer because they expect a number (Küchemann, 1981). To be able to cope with such a symbol requires not only that it be given a meaning, but that the meaning should allow the student the versatility of thought to see it as a procept, representing both as a process (of evaluation when x is known) and also an object that can be operated on. Often students who are used to working in the symbolic actions and symbolic process cells of FAMT see the symbol $x + 3$ solely as a process and not as a mental object; further it is a process they cannot carry out because they do not know what x is.

In two previous papers we have described (Graham & Thomas, 2000; Tall & Thomas, 1991) how DT might be used to help students construct a versatile perspective on the use of letters as generalised number. The basis of the approach used was to use digital technology to give students a symbolisation enabling an embodied view of the use of letter. This embodied, enactive perspective comprises a store with a label and a value that can be changed, as seen in Fig. 10.2, which is

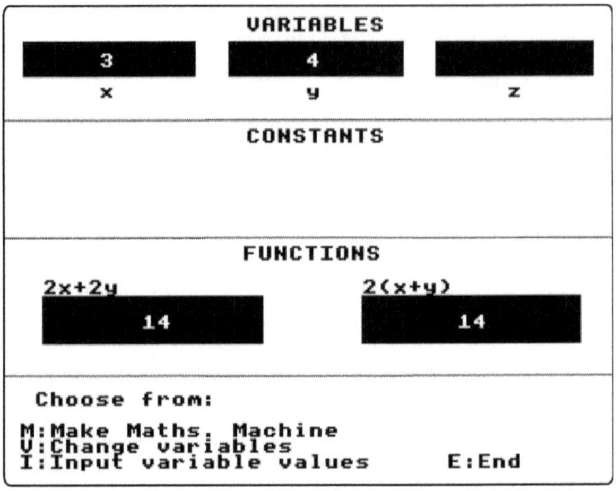

Fig. 10.2 The embodied symbolisation of a variable in the 'Maths Machine'

You can use letters as stores for numbers. Try the following:

Press	See	Explanation
4 [STO▶] [ALPHA] A [ENTER]	4 → A 4	The value 4 is stored in A.
[CLEAR]		This clears the display.
[ALPHA] **A** [ENTER]	A 4	This confirms that the number stored in A *is* 4.

Fig. 10.3 An example of the layout in Graham and Thomas (2000) algebra module

taken from Tall and Thomas (1991). Here students can engage in embodied actions, entering numbers into variable stores, predicting outcomes about algebraic objects and testing these predictions. In a second paper (Graham & Thomas, 2000) we changed the technology from computers to graphic calculators, which intrinsically employ variables with a large number of inbuilt stores labelled by the use of capital letters and where the embodied actions of storing and retrieving numbers from these lettered stores provides a direct correspondence to letter use in early algebra. The same basic embodied model was used here, the graphic calculator's lettered stores as a model of a variable. Each store is represented by a box in which changing values of the variable come and go, and next to which sits its label. Figure 10.3 shows a brief early section from the module used.

Both controlled experiments showed that the students using the DT were more versatile in their thinking than the students following a traditional course. They were significantly better at interpreting symbols, demonstrated an improved understanding of the use of letters as specific unknown and generalised number and were more likely to think of expressions as objects, without losing any procedural facility.

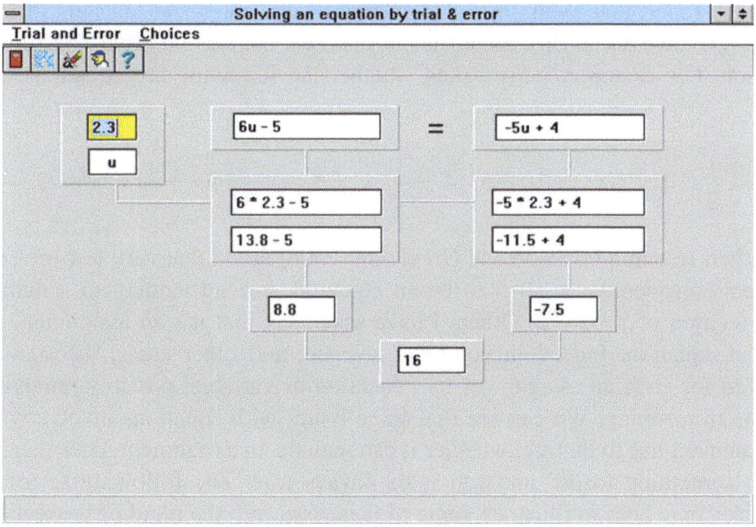

Fig. 10.4 A screen from the Dynamic Algebra programme

In terms of semiotics the research shows that by extending the sign or symbol used to represent a variable, from a single letter to a box plus a letter, students can be assisted to make improved sense of the object represented. Following from this approach a *Dynamic Algebra* programme was developed that enables investigative activity with expressions and equations based on the same mental model. In Fig. 10.4 we see an example of how this programme employs embodied actions such as giving the variable u a value to see the effect on two expressions $6u-5$ and $-5u+4$, to see when they reach equality. This is an example of an approach to the hardest type of linear equation at this level.

10.3 Equations

While the '=' sign is now ubiquitous in mathematics, making sense of the meaning of the sign appears not to be straightforward for students, and is often context dependent. For example, many have an operational, process-oriented perspective of the sign as a signal to perform some action (Crowley, Thomas, & Tall, 1994; Godfrey & Thomas, 2008; Kieran, 1981; Thomas, 1994). For these students there is a difference between, say,

$$2x+1=5 \text{ and } 3=5x+2$$
$$\frac{dy}{dx}=2x+5 \text{ and } 2x+\frac{dy}{dx}=3$$

I have found that even among mathematics graduates and teachers of mathematics we have some discussion in my master's courses on what constitutes an equation. For example, when asked whether the following are equations not all agree.

$$\frac{2x+1}{3x-2}=1,\ \ f(x)=2x+5,\ \ 4=4,\ \ k=5,\ \ (x-1)(x+3)=x^2+2x-3$$

In their research Hansson and Grevholm (2003) found that very few pre-service teachers considered $y=x+5$ to be an equation, instead tending to a numerical interpretation of $y=x+5$. Others I have asked say that it's an assignment rather than an equation. Indeed in computer science, and other areas, the sign := is reserved for such an assignment to a function or variable, possibly removing an overlap in meaning. We can see that some issues with equations involve whether the statement has to be true, whether it can include an assignment, does there have to be 'something to do' and can it be always true. The following set of three examples may help to illustrate some of these issues in the mind of the reader.

$$x^2+3x-1=x^2+3x+1,\quad (x-1)(x+3)=x^2+3x-3,\quad (x-1)(x+3)=x^2+2x-3$$

Addressing this the Collins mathematics dictionary (Borowski & Borwein, 1989) distinguishes between an *identical equation* (or identity), which is true for any values of the variables, and a *conditional equation*, which is only true for certain values of the variables. This distinction seems to be a useful one and it might help if more use were made of the symbol for equivalence (in an identical equation, true for all values of the variables), $\equiv$, that was more commonly used years ago.

In our study on understanding of equation (Godfrey & Thomas, 2008) we found that for Year 10 students (age 14–15 years) many appear to be using the criteria that an equation needs an = sign and an operation to carry out (see examples in Fig. 10.5). On this basis 65.6 % of them rejected $k=5$ as an equation while 72.4 % accepted $7w-w$ as an equation.

In this same study, for those in Year 13 (17–18 years old), the last year of school, 27.6 % still accepted $7w-w$ as an equation, while 56.6 % were unwilling to accept $a=5$ and 61.8 % did not see $a=a$ as an equation. Overall 53.9 % of these students still wanted an equation to have an operation to carry out, and 14.5 % of these

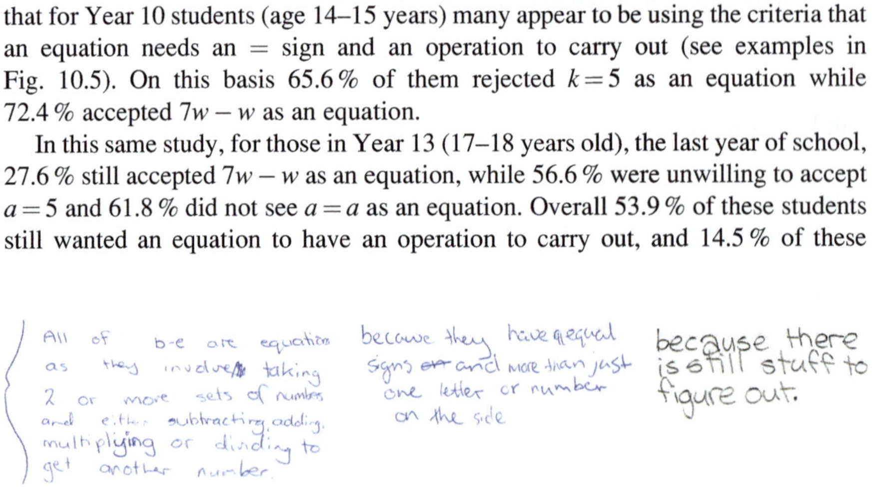

Fig. 10.5 Examples of 14-year-old students criteria for an equation

rejected anything that was an identity or an assignment. In our group of first year university students studying Engineering Science, which has a very high entry requirement, 20.6 % still emphasised the solution aspect of an equation (e.g. 'An equation is a mathematical formula formed by some unknown variables and numbers. And it is those unknown variables we are trying to find a value/answer to it' and 'Statement given to solve unknown variables in order to equate the right hand side is equal to the left hand side'). However, 60 % now accepted $a = a$ as an equation, although 26.7 % did not see $a + b = b + a$ as an equation.

Student understanding of the use of equality often appears not to be predicated on an explicit construction of properties of equations, including the reflexive, symmetric and transitive nature of the '=' sign, that will eventually lead to the idea of equivalence relations. Hence, activities that might allow them to construct some of these properties could be of value.

In addition what could we say to a student who produces this argument?

$$
\begin{aligned}
&4x^2 - 5x - 6 = 0\\
&(4x + 3)(x - 2) = 0\\
&4x + 3 = 0, \quad x - 2 = 0\\
&4x + 3 = x - 2 = 0\\
&4x + 3 = x - 2\\
&3x = -5\\
&x = -\frac{5}{3}
\end{aligned}
$$

Here the transitive property has been applied to $4x + 3 = 0,\ x - 2 = 0$ as if it reads $4x + 3 = 0$ *and* $x - 2 = 0$. Compare this with $a = b$ and $b = c$ implies $a = c$. However, the line actually should read $4x + 3 = 0$ *or* $x - 2 = 0$, and this might give a teacher the chance to discuss the important logical difference between 'and' and 'or' in mathematical statements. In this way a crucial link between symbolic algebra and logic using natural language could be made.

What about if we are working through an example where we are trying to find the intersection of two graphs, whose equations are $y_1 = x + 6$ and $y_2 = 3x + 1$? Is it necessary to explain how we get from line 1 to line 2 or how we have used the symmetric property that $a = b \Rightarrow b = a$ to get from line 4 to 5?

$$
\begin{aligned}
&y_1 = y_2\\
&x + 6 = 3x + 1\\
&6 = 2x + 1\\
&5 = 2x\\
&2x = 5\\
&x = 2\frac{1}{2}
\end{aligned}
$$

Or when solving $y_2 = 0$ and then using $3x + 1 = 0$ to do so, would we invoke the transitive property ($y_2 = 3x + 1$ and $y_2 = 0 \Rightarrow 3x + 1 = 0$)?

Equations of the type $ax + b = cx + d$ have been well known to be a cut-off point for those who will make good progress in the learning of algebra and the obstacle has been called the didactic cut (Filloy & Rojano, 1984) or cognitive gap (Herscovics & Linchevski, 1994). It is only at this point in the solving of equations that one has to operate on the variable. One approach when solving $ax + b = cx + d$, or similar equations such as those involving quadratic functions, is to assist student understanding of properties of equations, such as what a solution is and when it is invariant. For example, they might see that there is a difference between what I have called legitimate and productive transformations of an equation (see Hong, Thomas, & Kwon, 2000; Thomas, 2008a). A legitimate transformation of a linear equation adds $\pm k$ or $\pm kx$ for all real k to both sides, but a productive transformation that moves one quickly towards an algebraic solution, is one of the type $\pm ax$, $\pm cx$, $\pm b$, and $\pm d$, taken from the infinite number of legitimate transformations. It is important to understand that the solution remains invariant under both types of transformations. It may be that DT could be employed to help students see some properties of equations by linking the algebraic representation to the graphical one. Clearly adding $\pm k$ to both sides of the equation does not change the solutions because graphically we are translating both graphs parallel to the y-axis by $\pm k$. However, the effect of adding $\pm kx$ to both sides may not be so obvious. In Fig. 10.6a, which was constructed using GeoGebra, we can see that adding $\pm kx$ to both sides of the equation $2x + 2 = 5x - 3$ appears to rotate the graphs of the function on either side of the equation about the point of intersection with the y-axis, although the x-value of the point of intersection, the solution of the equation, remains invariant.

The angle a straight line $y = mx + c$ makes with the x-axis is given by $\tan\theta = m$, where θ is the angle with the x-axis, and adding kx will change it from $\theta = \tan^{-1}(m)$ to $\theta = \tan^{-1}(m + k)$, which may appear to indicate a rotational effect. However, while the angle the line makes with the horizontal changes the individual points do not rotate. Instead, in a move that encourages versatile thinking, we might utilise another area of mathematics; one that is sometimes less often employed in school mathematics, although it is essential for university studies in mathematics. The idea of a transformation of the plane represented in matrix form is very useful here. Linking mathematical ideas across representations in this way is very important (Duval, 2006) and is a way to promote representational versatility (Thomas, 2002, 2008a, 2008b). In essence adding kx to $f(x) = mx + c$ is a shear of the graph of the function by a factor k parallel to the y-axis. Using matrices and vectors we can represent this linear transformation as follows:

$$\begin{pmatrix} 1 & 0 \\ k & 1 \end{pmatrix}\begin{pmatrix} x \\ f(x) \end{pmatrix} = \begin{pmatrix} x \\ f(x) + kx \end{pmatrix} = \begin{pmatrix} x \\ mx + c + kx \end{pmatrix} = \begin{pmatrix} x \\ (m + k)x + c \end{pmatrix}$$

Now, since every point on the straight line (and in the plane), apart from those on the y-axis, which are all invariant, is moved parallel to the y-axis (giving the

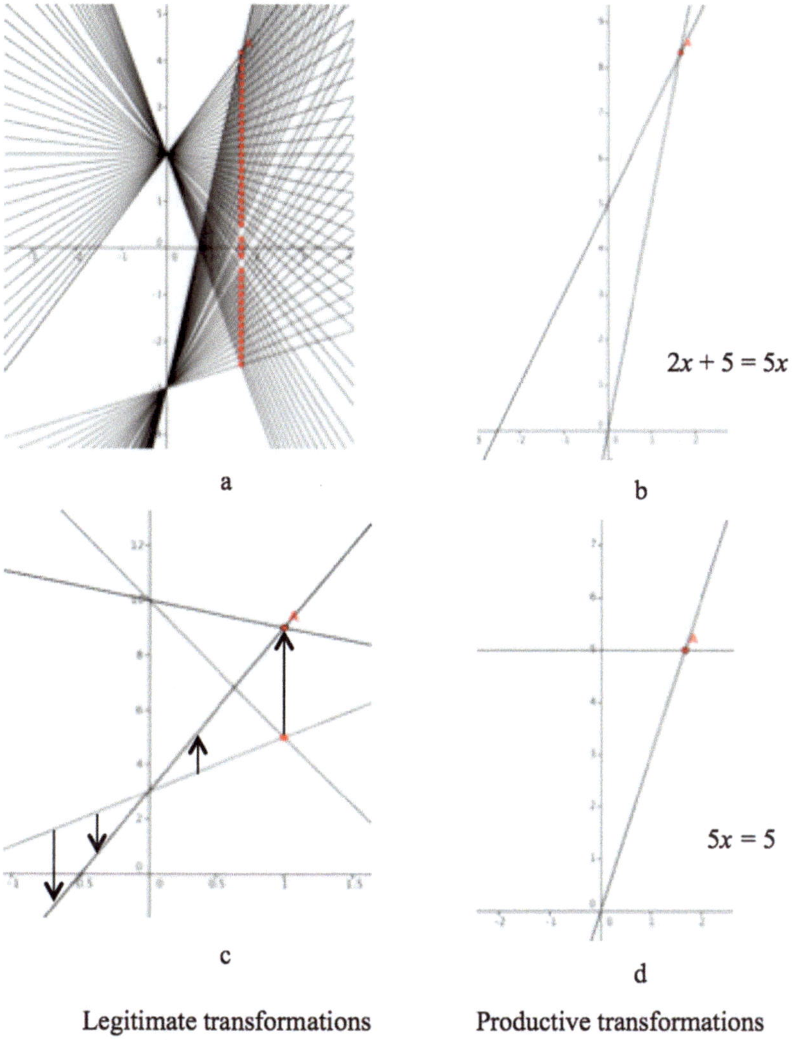

Fig. 10.6 Legitimate and productive transformations of linear equations

appearance of a rotation), we can see that this is also true of the point of intersection of two straight lines. We see this in Fig. 10.6c (we assume $k > 0$ here), and note that when $x < 0$ the points move in the opposite direction, since for $k > 0$, $kx < 0$. Thus the point of intersection ends up with the same x value as before, our invariant solution to the equation. In terms of the FAMT this process has linked a symbolic algebra process with an embodied graphical process and a symbolic matrix process. Further, we have managed to link a pointwise approach to a translation to a global perspective (Hong & Thomas, 2014; Vandebrouck, 2011).

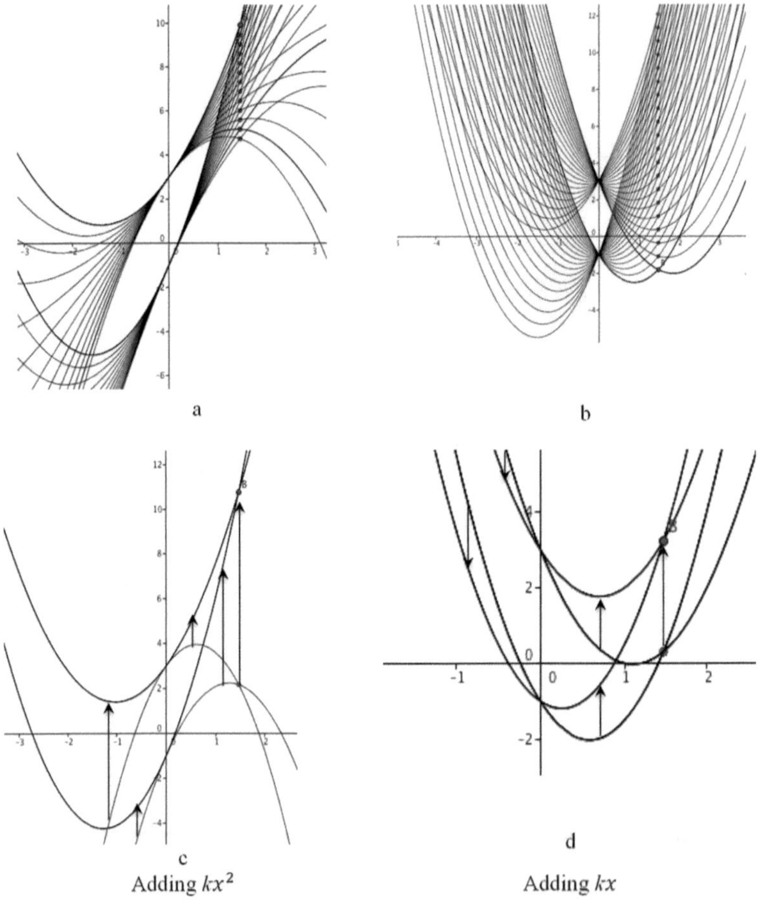

Fig. 10.7 Invariant solution under legitimate transformations of quadratic equations

Having established these basic principles we could now consider what happens with quadratic equations. Of course, the cases of adding a constant and adding a multiple kx of x can be analysed in exactly the same way as above, although the picture, again from GeoGebra, is quite different (see Fig. 10.7a, b). We can see that the case of adding kx^2 to both functions (see Fig. 10.7c) can be viewed in a similar manner to that of adding kx. The translation is again parallel to the y-axis and for $k > 0$, $kx^2 > 0$. Once again the point of intersection remains on each graph, the y-translation is by the same amount and the x-value is unchanged by adding to the value of the function. Hence, the solution is invariant.

10.4 Polynomial Functions

The concept of function, one of the most fundamental ideas in the whole of mathematics, is often given only cursory attention in school mathematics. Hence, it is not surprising that research has shown that students' perspectives on function differ considerably from those of mathematicians. For example, Williams (1998) used function concept maps to compare conceptions of students and professors and found that the students emphasised minor details and the idea that functions are equations. In contrast none of the professors thought of a function as an equation, preferring the idea of a correspondence, a mapping, a pairing or a rule. In a study with trainee mathematics teachers Chinnappan and Thomas (2003) found the teachers had a strong tendency to think of functions graphically and procedurally, and often even separated algebra from functions in their thinking. In Fig. 10.8 we see how a teacher, unable to decide on whether an ordered pair could represent a function, moves from the ordered pair representation to a graph and then to an explicit algebraic formula in order to say that this is a function.

The expectation that a function will have an explicit algebraic formula was prominent in Thomas' (2003) study. In Fig. 10.9a we see an example of how one

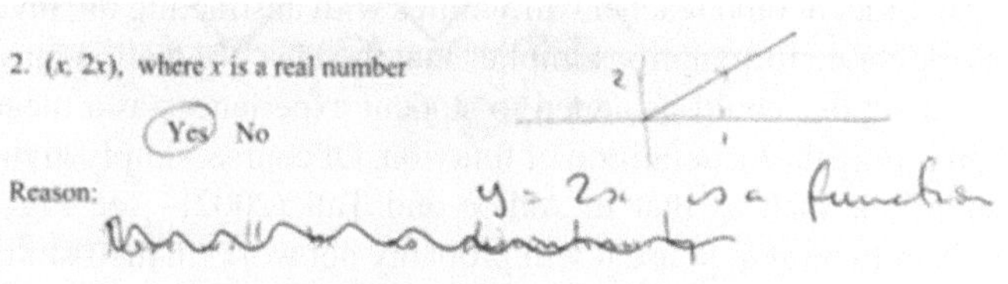

Fig. 10.8 A teacher's use of a graph and an algebraic formula for a function

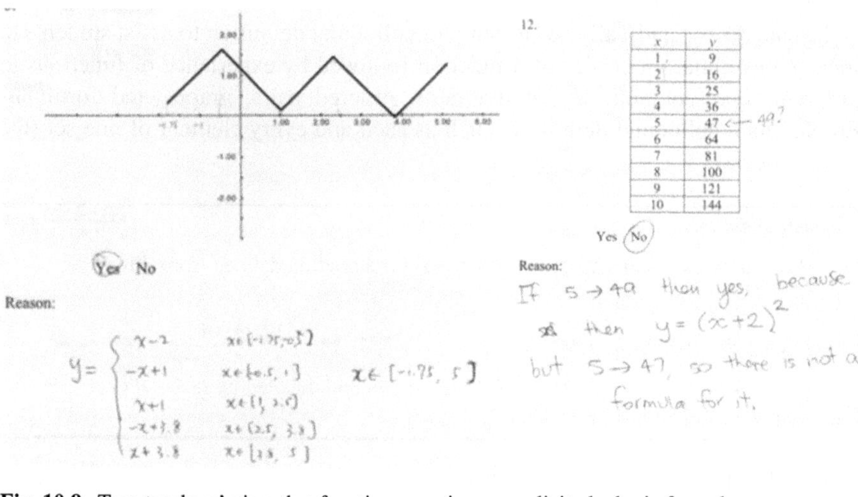

Fig. 10.9 Two teachers' view that functions require an explicit algebraic formula

teacher responded to the question of whether the given graph could represent a function by finding the explicit algebraic formula for each straight line section of the graph in order to be able to respond 'yes'. The second example in Fig. 10.9b shows the reverse. A teacher rejects the table of values as representing a function because the value at $x=5$ deviates from the formula $y=(x+2)^2$ that all the others fit.

This research suggests that for many teachers the graphical representation of function can become so dominant in thinking about function that it could hinder a growth in inter-representational understanding.

In terms of the FAMT framework it would appear that at least some students have a tendency to move between the embodied and symbolic worlds with respect to function. An emphasis on symbolic actions and processes may be behind the desire for an explicit formula and the use of the vertical line test embodied action/process may encourage a graphical perspective on function. This movement between embodied actions and symbolic actions is generally to be encouraged but abstracting the notion of a function from graphical and algebraic expressions exemplars appears to be difficult (Akkoc & Tall, 2002). As Thompson (1994) has pointed out, 'the core concept of 'function' is not represented by any of what are commonly called the multiple representations of function, but instead our making connections among representational activities produces a subjective sense of invariance' (p. 39). Student (and teacher) difficulties with abstracting the invariance from graphs and algebraic formulations implies that the idea of function may be one area where formal actions could be added to student experiences as a means of testing given constructs against a definition of function. Of course, simply giving students a formal definition, such as that in Akkoc and Tall (2002)—see Fig. 10.10—and expecting them to be able to use it will probably not work. In their study Akkoc and Tall (2002) found that some students were unable to see and apply the fundamental (simple) definition of function, instead relying on almost arbitrary aspects of examples they focussed on. Hence, the simplicity of the core function concept eluded most of their students.

Instead Akkoc and Tall used a four part colloquial definition to assist students to focus on essential properties of a function followed by experience of functions in different representations as set diagrams, ordered pairs, graphs and formulas. Employing a colloquial definition, such as each and every element of one set (the

Formal Definition

Let A and B two sets other than empty set, and f is a relation defined from A to B. If

i) $\forall x \in A$, $\exists y \in B$ s.t. $(x,y) \in f$,

ii) $(x,y_1) \in f$ and $(x,y_2) \in f \Rightarrow y_1 = y_2$

Then the relation f from A to B is called a *function* and denoted by $f: A \rightarrow B$ or $A \xrightarrow{f} B$.
A is called the domain and B is called the range.

Fig. 10.10 A possible formal world definition of function domain) is mapped to or related to one and only one element of the second set (or codomain) and then testing this with formal actions in the four representations used by Akkoc and Tall along with tables of values may be a way forward. Although as they found, this is not the complete solution.

These difficulties with thinking about the concept of function are further exemplified when students meet the idea of a polynomial. When asked what a polynomial is (see Chinnappan & Thomas, 2003) some trainee teachers responded:

- An equation which has more than 1 x variable whose power is bigger than 1
- An equation that has a power of x other than 1
- An equation with a power of x greater than one
- When I am talking about functions, I am not talking about polynomials and vice versa, I find it very difficult to um.. interchange the words
- If somebody said 'is that straight line relation a polynomial?', my gut reaction would be to say no. Just because a polynomial, poly being many.

So we can see an apparently common misconception here that linear functions, and by extension constant functions, are not polynomials and that the set of polynomials is not a subset of functions. Polynomials are perceived as beginning with the quadratic function, since that is probably where the term was first met. This view is reinforced by the natural language prefix 'poly', seen in other places in mathematics, such as polygon (where the number of sides has to be three or more). Confirming this are the kinds of responses received to the question of whether $3-x$ is a polynomial.

- No, linear
- No—The powers of x is low
- Yes—Not sure! Maybe it's not!
- Yes—because for each value of x, there is 1 corresponding y value

Once again, the idea of a polynomial (function) may be an area where it would be beneficial to add formal actions based on a definition to student experiences so that carefully chosen examples and non-examples of polynomials could be used to build the construct. For example, one could define a real polynomial of degree n as an expression of the form $p(x) = a_n x^n + a_{n-1}x^{n-1} + \cdots + a_2x^2 + a_1x + a_0$ where x is a real variable, n is a non-negative integer and each a_i is a real number (later we may define polynomials, for example, over the complex numbers) with $a_n \neq 0$. We can then use a formal action of testing against a definition to determine whether we have a polynomial or not, such as: Is $x+1$ a polynomial? Is $x^{5/2}$ a polynomial? Is 0 a polynomial?

When it comes to a consideration of the properties of some low order polynomials it would appear that, for cubic functions, a number of interesting areas for study have been often overlooked and would repay attention. I suggest one or two of these below that are accessible with DT.

10.5 Investigating Cubic Functions

One of the reasons for deciding what polynomial properties are studied in school algebra may be whether the properties are considered to be accessible to students through procedural calculations. However, with the advent of DT we can now investigate properties that may have previously been in the domain of 'higher' mathematics.

10.5.1 Symmetry

To simplify matters we will limit our discussion to monic cubic functions of the form $x^3 + ax^2 + bx + c$ with little loss of generality since $ax^3 + bx^2 + cx + d = a\left(x^3 + \frac{b}{a}x^2 + \frac{c}{a}x + \frac{d}{a}\right)$ when $a \neq 0$. For the function $x^3 + ax^2 + bx + c$ we note without proof here that the transformation $f\left(x - \frac{a}{3}\right)$ always removes the x^2 term (Why this works is an important question and CAS DT will confirm this). For example, if we have a function f with $f(x) = x^3 - 3x^2 + x - 5$, then $f\left(x - \frac{-3}{3}\right) = f(x+1) = (x+1)^3 - 3(x+1)^2 + (x+1) - 5$, which reduces to $x^3 - 2x - 6$. You might want to reach for your DT device to verify the above!

While this is an interesting property in its own right, it leads to two other interesting ideas. Firstly, if we draw the graph of the two functions, $f(x) = x^3 - 3x^2 + x - 5$ and $g(x) = x^3 - 2x - 6$, what do we find? Look at Fig. 10.11.

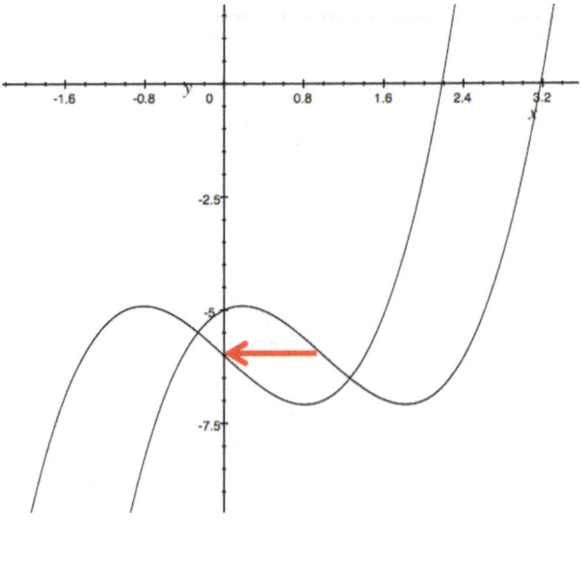

Fig. 10.11 An example of the graphical transformation of the cubic function for $f\left(x - \frac{a}{3}\right)$

Since $f'(x) = 3x^2 - 6x + 1$[1] and $f''(x) = 6x - 6 = 6(x-1)$ the cubic has a point of inflection at $x = 1$ and since $f(x+1)$ represents a translation of –1 parallel to the x-axis the point of inflection (1, –6) is mapped to (0, –6), on the y-axis. In general the point of inflection for the function j, where $j(x) = x^3 + ax + b$, will be mapped to (0, b). Looking at this transformation in general we note that for $p(x) = x^3 + ax^2 + bx + c$, $p''(x) = 2(3x + a)$, giving a point of inflection at $x = -\frac{a}{3}$. Hence, all cubic graphs have a point of inflection and the translation $p\left(x - \frac{a}{3}\right)$ moves the point of inflection to the y-axis. Of course, if $a = 0$ there is no x^2 term and the point of inflection is already on the axis.

Turning back to the function f we can move the point of inflection to the origin by adding 6 to $g(x) = x^3 - 2x - 6$, giving the function h, where $h(x) = x^3 - 2x$. Clearly h is an odd function (since $h(-x) = -h(x)$ for all x) and hence h has 180° rotational symmetry about the origin. The point is that this whole process generalises, so that translating $j(x) = x^3 + ax + b$ by $\frac{a}{3}$ parallel to the x-axis and then by $p\left(-\frac{a}{3}\right) = \dfrac{2a^3 - 9ab + 27c}{27}$ parallel to the y-axis the graph's point of inflection will be moved to the origin. Hence, we always end up with the odd function $x^3 - \dfrac{(a^2 - 3b)}{3}x$, showing that all cubic polynomials have rotational symmetry of 180° about the point of inflection $\left(-\dfrac{a}{3}, \dfrac{2a^3 - 9ab + 27c}{27}\right)$. Finding this general property can be made accessible to some students with the assistance of DT, as seen in Fig. 10.12, which was produced using TI-Nspire software.

Fig. 10.12 Using TI-Nspire software to show cubic symmetry

[1] In this chapter we make some use of calculus differentiation techniques. While calculus is usually not studied in school in the USA, many countries do include it in the curriculum from age 16 or 17. Since the primary aim of school algebra is to lead to calculus some minimal use seems reasonable.

10.5.2 Solving Equations

Interestingly the first step above, removing the term in x^2, was also the first step in the Tartaglia-Cardano method for solving cubic equations. If we then consider solutions to equations of the form $x^3 + ax + b = 0$, using Vieta's substitution, $x = z - \frac{a}{3z}$ enables us to solve the equation. For ease of calculation, although this is not crucial with DT, consider the equation $x^3 + 9x + 8 = 0$, where a is divisible by 3. We make the substitution $x = z - \frac{3}{z}$ and this gives rise to a 'disguised' quadratic that can easily be solved for z and hence x is found using $x = z - \frac{3}{z}$. Once again we show this process in Fig. 10.13, using TI-Nspire software. There are some things to note here. In Fig. 10.14 we move representations and draw the graph of the function f where $f(x) = x^3 + 9x + 8$, noting that the point of inflection appears on the y-axis as expected. This enables us to ask whether there is only one real root to the equation. We are trying to avoid calculus in this discussion where possible, since it lies beyond the remit of school algebra in the USA (see footnote 1), but note that since $f'(x) = 3x^2 + 9 > 0$ for all x the function is strictly (or monotone) increasing and so there is only one zero and hence only one real root of our equation. Other possible questions worth considering are whether this method always works (and if not when does it fail) and how we might find the complex roots.

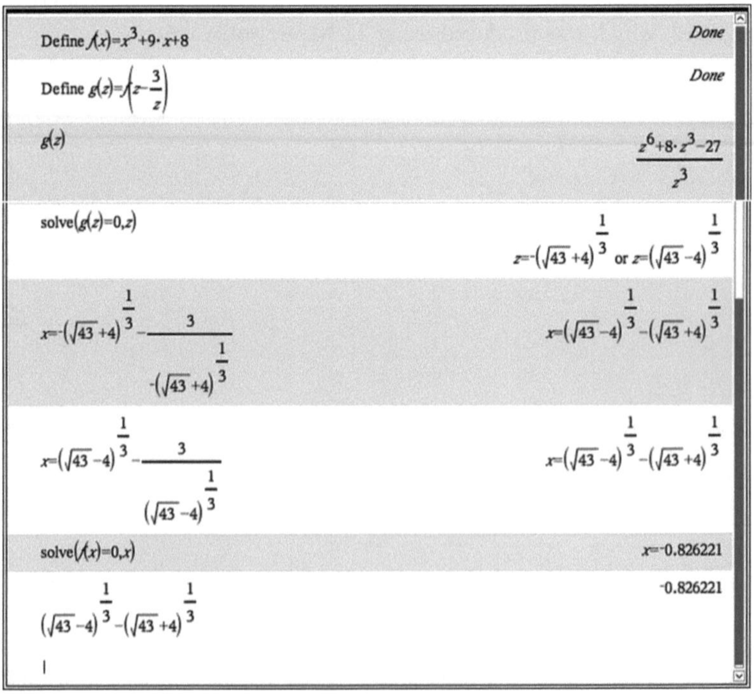

Fig. 10.13 Using TI-Nspire software to find exact solutions of cubic equations

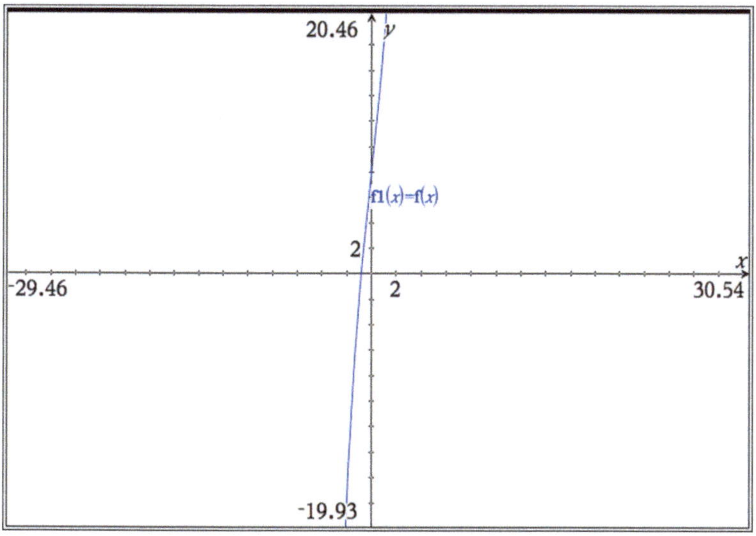

Fig. 10.14 Using TI-Nspire software to graph a solution to a cubic equation

10.5.3 Touching Graphs

Another task involving polynomials that could be given to students is:

Can we find quadratic functions whose graphs touch at a given point (p, q) with gradient k? How many possible graphs are there? Is there a general solution to the problem?

This task involves polynomials of degree 2 and links algebraic and graphical representations. While students can relatively easily find simple solutions, such as the graphs of polynomials x^2 and $-x^2$ that meet at (0, 0) with gradient 0, it is not so easy to solve more general cases by trial and error. However, once again this task is more approachable with DT. If we take a general quadratic function $f(x) = ax^2 + bx + c$ then we require the graph to pass through (p, q) and the gradient of the graph of the function at that point to be k. These two conditions can be written:

$$f(p) = ap^2 + bp + c = q \quad \text{and} \quad f'(p) = 2ap + b = k$$

In Fig. 10.15 we see the TI-Nspire software again employed to solve these equations simultaneously. The solution here is given in terms of a parameter c and, of course, p, q and k.

Choosing values for the point (p, q) and the gradient k gives a and b in terms of c, and we note that $c + kp - q \neq 0$ (since then we don't have a quadratic function) and $p \neq 0$. Figure 10.16 shows some of the possible solutions for the point (2, 3) and gradient 3 drawn using TI-Nspire. It is good practice to check these solutions, of course. For example, with $p = 2, q = 3, \ k = 1$ if we choose $c = -2$ then our function

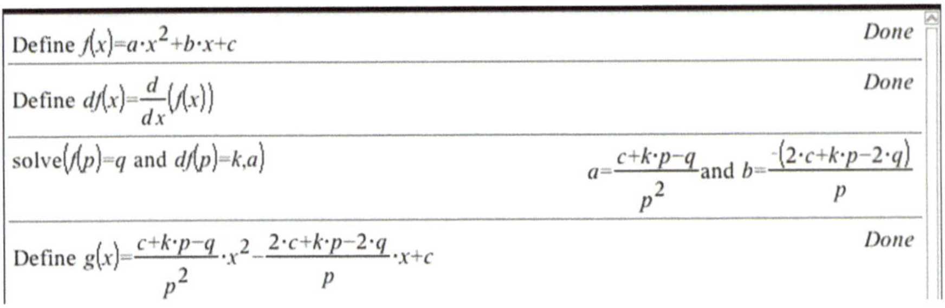

Fig. 10.15 Using the TI-Nspire software to find a general solution

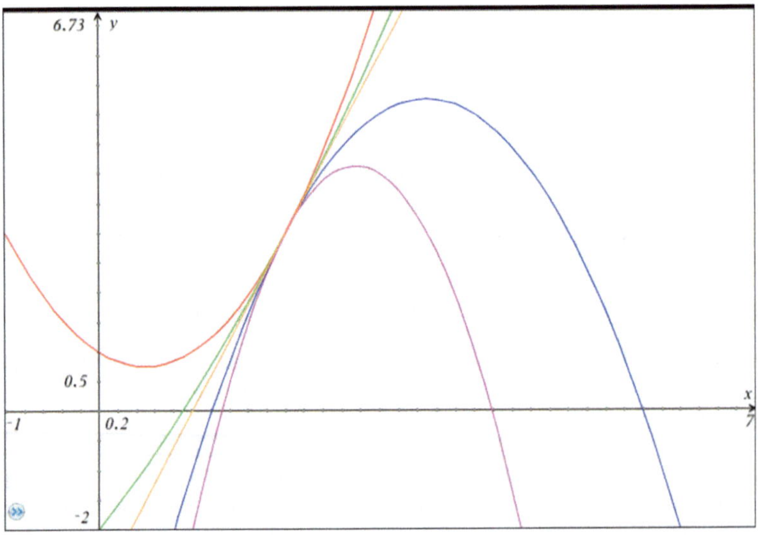

Fig. 10.16 Using the TI-Nspire software to show graphs of possible solution functions

is given by $f(x)=\frac{1}{4}x^2+2x-2$, which passes through (2, 3) and the gradient there is $\frac{1}{2}(2)+2=3$, as required.

A further question for investigation that arises is: does the latter condition $p\neq 0$ for the general solution mean that it is not possible to find graphs that meet on the y-axis with the same gradient? Well we have already seen that x^2 and $-x^2$ meet at (0, 0) with gradient 0, and in general so does kx^2, $k\neq 0$, k real. But what about other points not at the origin and whose gradient at $x=0$ is not zero? Well it certainly appears to be possible to find some, as Fig. 10.17 shows, but students will have to engage with how we might find these solutions. It is hoped that ways to structure interesting tasks for students that promote understanding of properties of polynomials will become apparent.

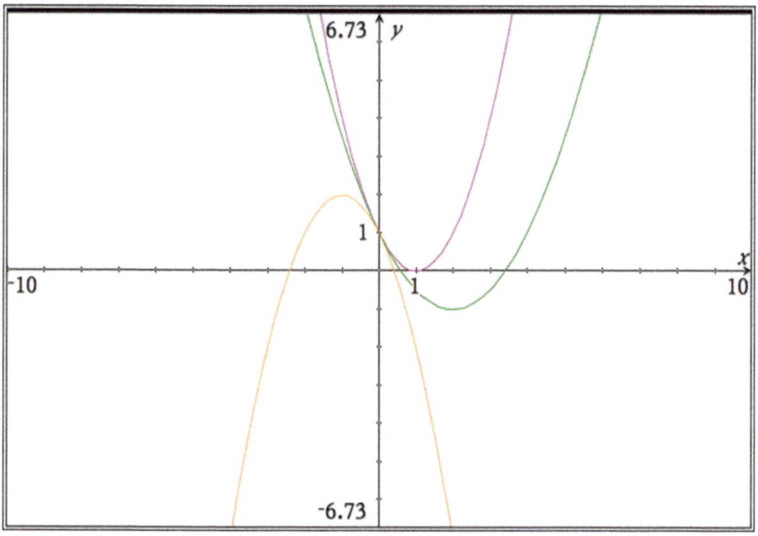

Fig. 10.17 Graphs of possible solution functions with $p = 0$

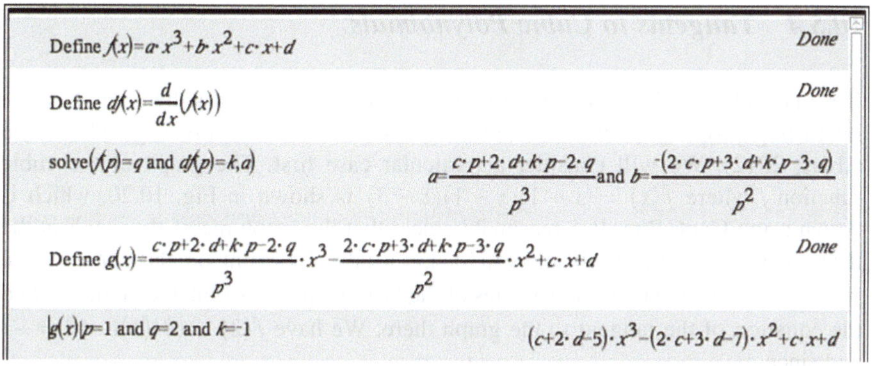

Fig. 10.18 Using the TI-Nspire software to find the general solution for cubic polynomials

Since generalising is always a key aim in mathematics, a possible next step is to try to extend these ideas further. One question we might ask is: Can we do the same for cubic polynomial functions? Using the DT again, as Fig. 10.18 shows, two parameters, c and d, are needed, where $(c + k)p + 2(d - q) \neq 0$ (since then we don't have a cubic) and $p \neq 0$, and Fig. 10.19 shows examples of the graphs of some polynomials of degree three meeting at the point (1, 2) with gradient –1 ($c + 2d - 5 \neq 0$ here).

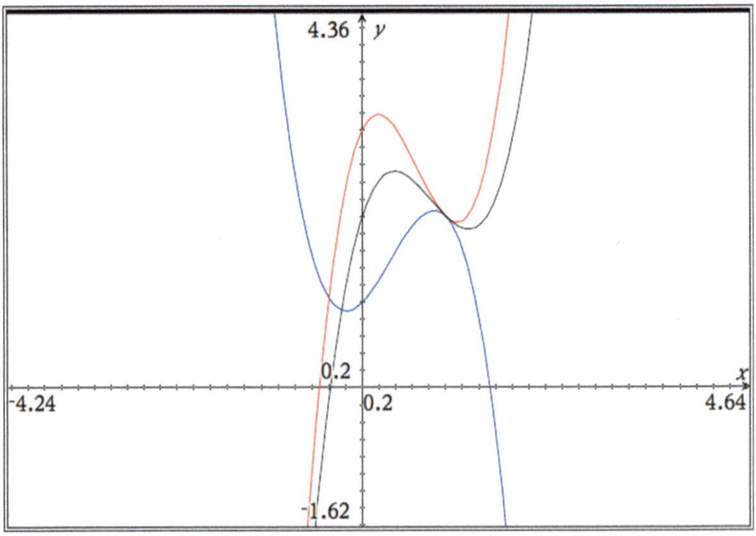

Fig. 10.19 Using the TI-Nspire software to show graphs of possible solution functions

10.5.4 Tangents to Cubic Polynomials

If we consider a cubic polynomial with three distinct real zeros then they have an interesting property related to their tangents that could be investigated (see de Alwis, 2012). We will consider a particular case first. The graph of the cubic function f where $f(x) = (x + 1)(x - 1)(x - 3)$ is shown in Fig. 10.20, which is drawn using GeoGebra. It is reasonably clear that the graph meets the x-axis at the three points (–1, 0), (1, 0) and (3, 0). Let's take the point on the curve where $x = \frac{1+3}{2} = 2$, the mean of the x values of the last two points of intersection, and find the equation of the tangent to the graph there. We have $f(2) = (3)(1)(-1) = -3$ and since

$$f(x) = (x + 1)(x - 1)(x - 3) = (x^2 - 1)(x - 3) = x^3 - 3x^2 - x + 3$$
$$f'(x) = 3x^2 - 6x - 1$$

and $f'(2) = -1$. So the equation of the tangent is $y + 3 = -1(x - 2)$ or $y + x + 1 = 0$ and when $y = 0$ for this tangent $x = -1$. So the tangent at the mean value of two points of intersection passes through the third point of intersection. Figure 10.20 also shows the tangent at the point where $x = \frac{-1+1}{2} = 0$ passing through the point (3, 0).

Of course, the tangent at the point where $x = \frac{-1+3}{2} = 1$ passes through the point (1, 0) here since it's a special case where the zeros are equally spaced. So the question is does this result generalise? Is it always true for cubics? One way to investigate it using GeoGebra is to use sliders for the function coefficients.

Fig. 10.20 The graph of the cubic function

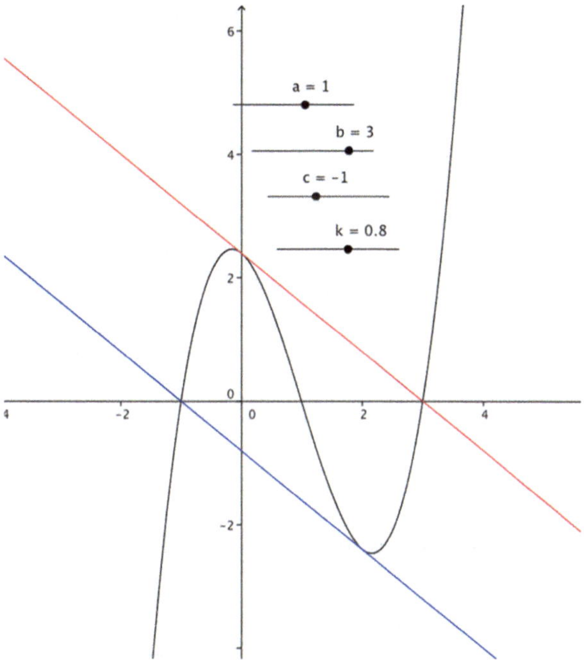

$f(x) = (x + 1)(x - 1)(x - 3)$

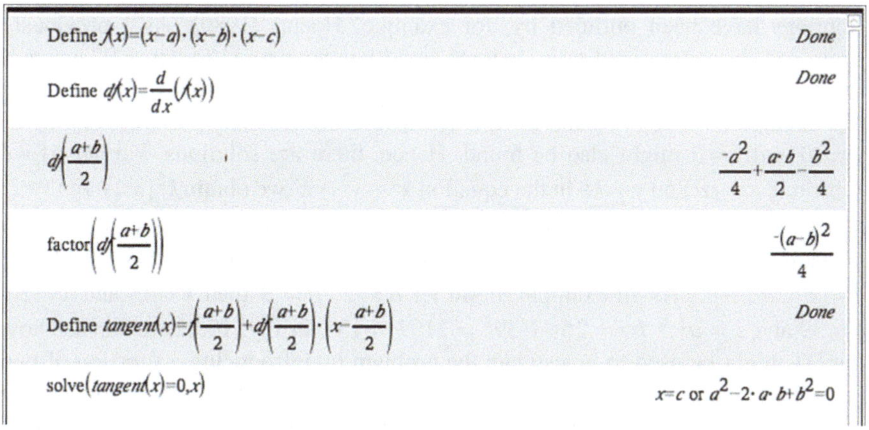

Fig. 10.21 Using TI-Nspire to demonstrate the generality of the tangent property of the cubic function $f(x) = (x - a)(x - b)(x - c)$

Algebraically, consider the monic polynomial function f where $f(x) = (x - a)(x - b)(x - c)$, and without loss of generality consider the tangent at the point M where $x = \frac{a+b}{2}$. This could be done by hand but once again the symbolic process can be left to the DT, in this with case TI-Nspire as shown in Fig. 10.21. The function *df* is the derivative of *f* and we note that the DT does not automatically factorise the result, although this is not crucial in this example. Using df we can find the gradient of the tangent at the point where $x = \frac{a+b}{2}$ and hence the equation of the tangent using the well-known equation $y - y_1 = m(x - x_1)$, where m is the gradient and (x_1, y_1) a point on the line (nb $y = y_1 + m(x - x_1)$ is used here). Then solving for where the tangent is zero gives $x = c$ or, interestingly, $a^2 - 2ab + b^2 = 0$ but then $(a - b)^2 = 0,\ a = b$, which would contradict our requirement that f have three distinct real zeros. So the tangent at $x = \frac{a+b}{2}$ does indeed pass through the point $(c, 0)$.

Once more the DT has allowed us to make some crucial links, this time between embodied actions and processes involving graphs and tangents and symbolic processes in order to find a solution for the task. Often we make the link by encouraging embodied views of symbolic expressions, so it is good to have an example that links the representations in the other direction.

10.6 Polynomials in Two Variables

Students at school often consider Pythagoras' theorem and its solutions, and while the theorem does not generalise to higher powers, as Fermat's last theorem states, solutions to other Diophantine equations are in reach if we use DT. One of these that can be approached, that I have described elsewhere (see Heid, Thomas, & Zbiek, 2013), is $x^2 + y^2 = z^3$, a special case of the general equation $x^n + y^n = z^{n+1}$, whose solutions have been outlined by, for example, Hoehn (1989). As I previously suggested, in a structured task students could be encouraged to use a DT spreadsheet listing values of n^2 and n^3 to try to find two of the squares that add up to a cube (for example, $x = 2$, $y = 2$ and $z = 2$ may be seen immediately). In this way $x = 5$, $y = 10$ and $z = 5$ might also be found. Hence, there are solutions. Further, if we substitute $x = ka$ and $y = kb$ in the equation $x^2 + y^2 = z^3$ we obtain $k^2\left(a^2 + b^2\right) = z^3$ and although this substitution is not obvious this last equation gives a big leap forward to finding solutions, since setting $k = a^2 + b^2$ will produce a solution $z = k = a^2 + b^2$. As an example, if we let $a = 2,\ b = 3$ then $k = 13$ and $x = 26$, $y = 39$ and $z = a^2 + b^2 = 26^2 + 39^2 = 2197 = 13^3$. In Fig. 10.22 we can see how the DT might be used to investigate the problem by introducing a function of two variables (we can also see this as a polynomial in two variables), an idea that will be very important later in mathematics. Hence, this constitutes an example of mathematics at the horizon in the mathematical knowledge for teaching framework (Ball, Hill, & Bass, 2005; Hill & Ball, 2004).

Extending the same method to a general equation $x^n + y^n = z^{n+1}$ could be too difficult for most school students, but the method above does generalise and this can be seen using DT, as in Fig. 10.23. Interestingly, as shown, the factorisation of $(a(a^n + b^n))^n + (b(a^n + b^n))^n$ seems beyond this DT program, but those students who have been taught to 'see' algebraic factors may be able to work as follows:

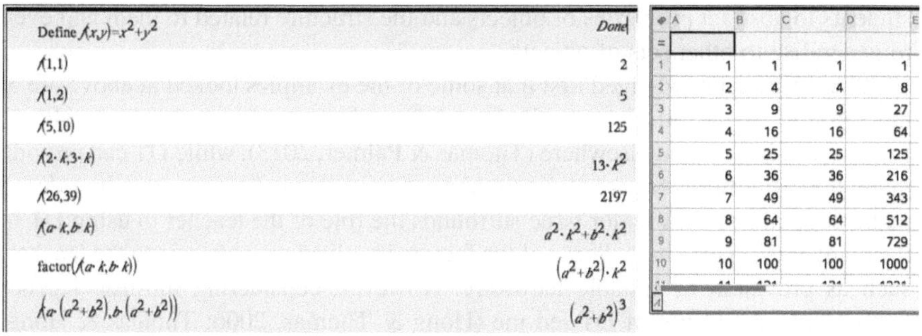

Fig. 10.22 Linking representations to find solutions to $x^2 + y^2 = z^3$

Fig. 10.23 Using DT to find solutions to $x^n + y^n = z^{n+1}$

$$\begin{aligned}(a(a^n + b^n))^n + (b(a^n + b^n))^n &= a^n(a^n + b^n)^n + b^n(a^n + b^n)^n \\ &= (a^n + b^n)^n\{a^n + b^n\} = (a^n + b^n)^{n+1}\end{aligned}$$

and hence this leads to a solution with $x = a(a^n + b^n)$, $y = b(a^n + b^n)$ and $z = a^n + b^n$.

10.7 Concluding Remarks

In a standard algebra curriculum students are involved in a great deal of what we have called manipulation algebra (Thomas & Tall, 2001). The outcome of this practice is that students may learn a lot about manipulating symbolic literals but far less about the nature of the objects they represent, such as polynomial functions, and their properties. Stressing the value of enactive and iconic thinking (Bruner, 1966) through visualisation encourages students to engage in the inter-representational conversions (Duval, 2006) that are a crucial constituent of building versatile thinking. Central to that inter-representational thinking is the DT, which, if it is used thoughtfully, can take on the role of epistemic mediator in order to help students to abstract properties of objects and the structure related to them and even to generalise to other sets of objects.

It has to be acknowledged first that some of the examples looked at above are at the top end of the difficulty scale for students in secondary school or college. Further, as I have noted elsewhere (Thomas & Palmer, 2013), while DT can provide many opportunities there are also a number of obstacles to be overcome in order to make good use of it. A major issue surrounds the role of the teacher in using DT in the manner described here. Some of the factors involved are extrinsic to the teacher, such as provision of suitable hardware. However, considering intrinsic teacher factors influencing use of DT led me (Hong & Thomas, 2006; Thomas & Hong, 2005) to propose an emerging framework for *pedagogical technology knowledge* (PTK) as a construct that could be an indicator of teacher progress in implementation of technology use. A teacher's PTK incorporates the principles, conventions, and techniques required to teach mathematics through DT. While the teacher has to be a proficient user of the technology she must also understand what is required to build tasks and situations that incorporate it, in order to enable mathematical learning through the technology. The essential teacher factors that combine to produce PTK include: instrumental genesis; mathematical knowledge for teaching; orientations and goals (Schoenfeld, 2011), especially beliefs about the value of technology and the nature of learning mathematical knowledge; and other affective aspects, such as confidence in teaching with DT.

In spite of these reservations I suggest that a rethink of the algebra curriculum and the dominance of the symbol manipulation approach usually employed might pay dividends in terms of stimulating versatile thinking by students and hence improve understanding of algebra.

References

Akkoc, H., & Tall, D. O. (2002). The simplicity, complexity and complication of the function concept. In A. D. Cockburn & E. Nardi (Eds.), *Proceedings of the 26th Conference of the International Group for the Psychology of Mathematics Education* (Vol. 2, pp. 25–32). Norwich, UK.

Ball, D. L., Hill, H. C., & Bass, H. (2005). Knowing mathematics for teaching: Who knows mathematics well enough to teach third grade, and how can we decide? *American Educator, 29* (1), 14–17, 20–22, 43–46.

Borowski, E. J., & Borwein, J. M. (1989). *Dictionary of mathematics*. London: Collins.

Bruner, J. (1966). *Toward a theory of instruction*. Cambridge, MA: Harvard University Press.

Chinnappan, M., & Thomas, M. O. J. (2003). Teachers' function schemas and their role in modelling. *Mathematics Education Research Journal, 15*(2), 151–170.

Crowley, L., Thomas, M. O. J., & Tall, D. O. (1994). Algebra, symbols and translation of meaning. In J. P. da Ponte & J. F. Matos (Eds.), *Proceedings of the 18th Conference of the International Group for the Psychology of Mathematics Education* (Vol. 2, pp. 240–247). Lisbon, Portugal: Program Committee.

de Alwis, A. (2012). Some curious properties and loci problems associated with cubics and other polynomials. *International Journal of Mathematical Education in Science and Technology, 43* (7), 897–910.

Dubinsky, E. (1991). Reflective abstraction in advanced mathematical thinking. In D. O. Tall (Ed.), *Advanced mathematical thinking* (pp. 95–123). Dordrecht: Kluwer Academic.

Dubinsky, E., & McDonald, M. (2001). APOS: A constructivist theory of learning. In D. Holton (Ed.), *The teaching and learning of mathematics at university level: An ICMI study* (pp. 275–282). Dordrecht, The Netherlands: Kluwer Academic.

Duval, R. (2006). A cognitive analysis of problems of comprehension in a learning of mathematics. *Educational Studies in Mathematics, 61*, 103–131.

Ernest, P. (2006). A semiotic perspective of mathematical activity: The case of number. *Educational Studies in Mathematics, 61*, 67–101.

Filloy, E., & Rojano, T. (1984). From an arithmetical to an algebraic thought. In J. M. Moser (Ed.), *Proceedings of the Sixth Annual Meeting of PME-NA* (pp. 51–56). Madison: University of Wisconsin.

Godfrey, D., & Thomas, M. O. J. (2008). Student perspectives on equation: The transition from school to university. *Mathematics Education Research Journal, 20*(2), 71–92.

Graham, A. T., Pfannkuch, M., & Thomas, M. O. J. (2009). Versatile thinking and the learning of statistical concepts. *ZDM: The International Journal on Mathematics Education, 45*(2), 681–695.

Graham, A. T., & Thomas, M. O. J. (2000). Building a versatile understanding of algebraic variables with a graphic calculator. *Educational Studies in Mathematics, 41*(3), 265–282.

Graham, A. T., & Thomas, M. O. J. (2005). Representational versatility in learning statistics. *International Journal of Technology in Mathematical Education, 12*(1), 3–14.

Hansson, O., & Grevholm, B. (2003). Preservice teachers' conceptions about $y = x + 5$: Do they see a function? *Proceedings of the 27th Conference of the International Group for the Psychology of Mathematics Education* (Vol. 3, pp. 25–32). Honolulu, Hawaii.

Heid, M. K., Thomas, M. O. J., & Zbiek, R. M. (2013). How might computer algebra systems change the role of algebra in the school curriculum? In A. J. Bishop, M. A. Clements, C. Keitel, J. Kilpatrick, & F. K. S. Leung (Eds.), *Third international handbook of mathematics education* (pp. 597–642). Dordrecht: Springer.

Herscovics, N., & Linchevski, L. (1994). A cognitive gap between arithmetic and algebra. *Educational Studies in Mathematics, 27*, 59–78.

Hill, H., & Ball, D. L. (2004). Learning mathematics for teaching: Results from California's mathematics professional development institutes. *Journal for Research in Mathematics Education, 35*, 330–351. doi:10.2307/30034819.

Hodgen, J., Brown, M., Küchemann, D., & Coe, R. (2010). *Mathematical attainment of English secondary school students: A 30-year comparison*. Paper presented at the British Educational Research Association (BERA) Annual Conference, University of Warwick.

Hodgen, J., Coe, R., Brown, M., & Küchemann, D. E. (2014). Improving students' understanding of algebra and multiplicative reasoning: Did the ICCAMS intervention work? In S. Pope (Ed.), *Proceedings of the Eighth British Congress of Mathematics Education* (BCME8) (pp. 1–8). University of Nottingham.

Hoehn, L. (1989, December). Solutions of $x^n + y^n = z^{n+1}$. *Mathematics Magazine*, 342. doi:10.2307/2689491.

Hong, Y. Y., & Thomas, M. O. J. (2006). Factors influencing teacher integration of graphic calculators in teaching. In *Proceedings of the 11th Asian Technology Conference in Mathematics* (pp. 234–243). Hong Kong.

Hong, Y. Y., & Thomas, M. O. J. (2014). Graphical construction of a local perspective on differentiation and integration. *Mathematics Education Research Journal, 27*, 183–200. doi:10.1007/s13394-014-0135-6.

Hong, Y. Y., Thomas, M. O. J., & Kwon, O. (2000). Understanding linear algebraic equations via super-calculator representations. In T. Nakahara & M. Koyama (Eds.), *Proceedings of the 24th Annual Conference of the International Group for the Psychology of Mathematics Education* (Vol. 3, pp. 57–64). Hiroshima, Japan: Programme Committee.

Kieran, C. (1981). Concepts associated with the equality symbol. *Educational Studies in Mathematics, 12*(3), 317–326.

Küchemann, D. E. (1981). Algebra. In K. M. Hart (Ed.), *Children's understanding of mathematics: 11-16* (pp. 102–119). London: John Murray.

Mason, J. (1987). What do symbols represent? In C. Janvier (Ed.) *Problems of representation in the teaching and learning of mathematics*. Hillsdale, NJ: LEA.

Peirce, C. S. (1898). Logic as semiotic: The theory of signs. In J. Bucher (Ed.), *Philosophical writings of Peirce*. New York: Dover.

Rosnick, P., & Clement, J. (1980). Learning without understanding: The effect of tutorial strategies on algebra misconceptions. *Journal of Mathematical Behavior, 3*(1), 3–27.

Russell, B. (1903). *The principles of mathematics*. Cambridge: Cambridge University Press.

Schoenfeld, A. H. (2011). *How we think. A theory of goal-oriented decision making and its educational applications*. Routledge: New York.

Skemp, R. (1971). *The psychology of learning mathematics*. Middlesex, UK: Penguin.

Skemp, R. R. (1976). Relational understanding and instrumental understanding. *Mathematics Teaching, 77*, 20–26.

Stewart, S., & Thomas, M. O. J. (2010). Student learning of basis, span and linear independence in linear algebra. *International Journal of Mathematical Education in Science and Technology, 41*(2), 173–188.

Struik, D. J. (1969). *A source book in mathematics, 1200-1800*. Cambridge, MA: Harvard University Press.

Tall, D. O. (2004). Building theories: The three worlds of mathematics. *For the Learning of Mathematics, 24*(1), 29–32.

Tall, D. O. (2008). The transition to formal thinking in mathematics. *Mathematics Education Research Journal, 20*(2), 5–24.

Tall, D. O., & Thomas, M. O. J. (1991). Encouraging versatile thinking in algebra using the computer. *Educational Studies in Mathematics, 22*, 125–147.

Tall, D. O., Thomas, M. O. J., Davis, G., Gray, E., & Simpson, A. (2000). What is the object of the encapsulation of a process? *Journal of Mathematical Behavior, 18*(2), 223–241.

Thomas, M. O. J. (1988). *A conceptual approach to the early learning of algebra using a computer*. Unpublished PhD thesis, University of Warwick.

Thomas, M. O. J. (1994). A process-oriented preference in the writing of algebraic equations. In G. Bell, B. Wright, N. Leeson, & J. Geake (Eds.), *Challenges in mathematics education: Constraints on construction. Proceedings of the 17th Mathematics Education Research Group of Australasia Conference* (pp. 599–606). Lismore, Australia: MERGA.

Thomas, M. O. J. (2002). Versatile thinking in mathematics. In D. O. Tall & M. O. J. Thomas (Eds.), *Intelligence, learning and understanding in mathematics* (pp. 179–204). Flaxton, Queensland, Australia: Post Pressed.

Thomas, M. O. J. (2003). The role of representation in teacher understanding of function. In N. A. Pateman, B. J. Dougherty, & J. Zilliox (Eds.), *Proceedings of the 27th Conference of the International Group for the Psychology of Mathematics Education* (Vol. 4, pp. 291–298). Honolulu, Hawai'i: University of Hawai'i.

Thomas, M. O. J. (2008a). Conceptual representations and versatile mathematical thinking. *Proceedings of ICME-10* (CD version of proceedings). Copenhagen, Denmark, 1–18. Retrieved from http://www.icme10.dk/proceedings/pages/regular_pdf/RL_Mike_Thomas.pdf

Thomas, M. O. J. (2008b). Developing versatility in mathematical thinking. *Mediterranean Journal for Research in Mathematics Education, 7*(2), 67–87.

Thomas, M. O. J., & Hong, Y. Y. (2005). Teacher factors in integration of graphic calculators into mathematics learning. In H. L. Chick & J. L. Vincent (Eds.), *Proceedings of the 29th Conference of the International Group for the Psychology of Mathematics Education* (Vol. 4, pp. 257–264). Melbourne, Australia: University of Melbourne.

Thomas, M. O. J., & Palmer, J. (2013). Teaching with digital technology: Obstacles and opportunities. In A. Clark-Wilson, N. Sinclair, & O. Robutti (Eds.), *The mathematics teacher in the digital era* (pp. 71–89). Dordrecht: Springer.

Thomas, M. O. J., & Stewart, S. (2011). Eigenvalues and eigenvectors: Embodied, symbolic and formal thinking. *Mathematics Education Research Journal, 23*(3), 275–296. doi:10.1007/s13394-011-0016-1.

Thomas, M. O. J., & Tall, D. O. (2001). The long-term cognitive development of symbolic algebra. In *Proceedings of the International Congress of Mathematical Instruction (ICMI) the Future of the Teaching and Learning of Algebra* (pp. 590–597). Melbourne.

Thompson, P. W. (1994). Students, functions, and the undergraduate curriculum. In E. Dubinsky, A. Schoenfeld, & J. Kaput (Eds.), *Research in collegiate mathematics education, I, Issues in mathematics education* (Vol. 4, pp. 21–44). Providence, RI: American Mathematical Society.

Vandebrouck, F. (2011). Students' conceptions of functions at the transition between secondary school and university. In M. Pytlak, T. Rowland, & E. Swoboda (Eds.), *Proceedings of the 7th Conference of European Researchers in Mathematics Education* (pp. 2093–2102). Poland: Rzeszow.

Wagner, S. (1981). Conservation of equation and function under transformations of variable. *Journal for Research in Mathematics Education, 12*, 118–197.

Williams, C. G. (1998). Using concept maps to assess conceptual knowledge of function. *Journal for Research in Mathematics Education, 29*(4), 414–421.

Part V
Teaching Higher Algebra

Chapter 11
Why Does Linear Algebra Have to Be So Abstract?

John Hannah

Abstract Research has shown that students struggle with the abstraction of linear algebra and many remedies have been tried. Here I offer another idea to add to your arsenal. Instead of presenting linear algebra as a stand-alone subject, deduced logically from a founding set of axioms, maybe we could present it as a subject that evolves naturally from students' experiences, either from prior contact with vectors in a physics course, or else from discussions and experiments designed to provoke a need to abstract, to generalize, to define and to prove.

Keywords Vector • Linear independence • Computer-based experiments • Concreteness • Generalizability • Necessity • Need for proof

11.1 Introduction

One website about the applications of linear algebra gives, as its first application, Abstract Thinking and offers the following advice on the subject.

> *One thing you can learn from the definitions, theorems and proofs you'll see in Linear Algebra ... is how to think clearly and express yourself clearly, to avoid misunderstanding and confusion.* (*Khoury*, 2006)

Unfortunately, research in mathematics education has found that misunderstanding and confusion are often the hallmarks of students after a first course in linear algebra; see Dorier (2000) and Thomas et al. (2015) for overviews of research into the teaching of introductory linear algebra courses. New definitions, in particular, are so numerous that students are essentially learning a new language. Furthermore, this language is used to describe a world containing very few familiar points of reference. So students end up feeling lost. This is doubly unfortunate because the ideas of linear algebra are finding applications, not just in other branches of mathematics, but also in such diverse areas as engineering, physics, economics,

---

J. Hannah (✉)
School of Maths and Statistics, University of Canterbury, Private Bag 4800, Christchurch, New Zealand
e-mail: john.hannah@canterbury.ac.nz

S. Stewart (ed.), *And the Rest is Just Algebra*, DOI 10.1007/978-3-319-45053-7_11

image processing and genetics. How can we hope to open up access to linear algebra to students of these subjects when mathematics students struggle so much to get started?

It is over 20 years since the Linear Algebra Curriculum Study Group (LACSG) issued its recommendations for reforming the teaching of linear algebra (Carlson, Johnson, Lay, & Porter, 1993) and there is no doubt that textbooks have changed somewhat in that time, with more emphasis on matrices and applications, see Lay (2012) and Poole (2011), for example. However, as we shall see, the traditional exposition in terms of definitions, theorems and proofs has not changed very much at all. In his discussion of the LACSG recommendations, Harel (2000) recommended three principles to be taken into account when dealing with the abstraction of linear algebra:

- (Concreteness) If students are to abstract a mathematical structure from a given model of that structure, the elements of that model must be familiar conceptual entities in the students' eyes.
- (Necessity) If students are to learn something, they must see a need for what they are being taught.
- (Generalizability) When teaching with a 'concrete' model, that is, a model that satisfies the Concreteness principle, your instructional activities within this model should allow and encourage the generalizability of concepts.

This suggests a different approach to the abstract ideas of linear algebra, whereby concepts emerge from student experiences rather than from unmotivated axioms and definitions. In this chapter I will offer some examples of such an approach. The idea is that concepts and their definitions will arise naturally, as important themes or techniques recur and ask to be named. Students will learn about these concepts and their definitions by using them repeatedly, and will tease out their precise meaning by seeing them used in a variety of contexts, much as we all learned our mother tongue as children. This approach is similar in spirit to the ideas promoted by John Mason (see Chap. 6, this volume), where he suggests that beginners in algebra need to appreciate how equations emerge from an expression of generality, if they are to have any chance of understanding the role that algebra can play in solving those equations.

In Sect. 11.2 we see an example of how we can build on students' previous experience of vectors, usually in a physics course dealing with concepts such as force or velocity, integrating this with a more mathematical view of vectors as ordered n-tuples. Section 11.3 looks at how the concept of linear dependence might emerge naturally from discussions of a geometric problem in two or three dimensions. Although the limitations of two or three dimensions may restrict the generalizability of this concept, we shall see how computer-based experiments can be used break beyond those limitations and help students develop a more general intuition for linear dependence. In Sect. 11.4 we see how such experiments can also help students to see a need for proofs or counterexamples.

11.2 Why Aren't Vectors Always Arrows?

Most New Zealand students meet vectors for the first time as part of a physics course at high school. Here vectors are quantities which have both magnitude and direction (such as force or velocity) and can be represented by arrows (or directed line segments) in two or three dimensions. At my university, mathematics students (whether they are majoring in mathematics or in science or engineering) then meet matrices as part of an approach to solving systems of linear equations, and they are introduced to an apparently different kind of vector, namely, a matrix with just one column (or row). Thus, solutions to a system of linear equations can be represented in terms of such vectors, one component for each unknown, even if the system has more than three unknowns.

It is perhaps a matter of taste whether one gives priority to the 'physics' idea of a vector or to the 'mathematics' idea. Two recent editions of popular textbooks illustrate the range of possible attitudes. For example, Anton's introduction to calculus (Anton, Bivens, & Davis, 2012, Sections 11.2–11.4) tells students that vectors are quantities which have both magnitude and direction, represented geometrically by arrows or directed line segments. The dot product $\mathbf{u} \cdot \mathbf{v}$ and cross product $\mathbf{u} \times \mathbf{v}$ are then defined ('mathematically', we might say) in terms of the components of the vectors **u** and **v**, with their usual 'physics' definitions (in terms of $|\mathbf{u}|\,|\mathbf{v}| \cos\theta$ and $|\mathbf{u}|\,|\mathbf{v}| \sin\theta$) being derived and then used to calculate physical quantities such as components or projections, or geometric quantities such as areas or volumes, associated with arrows or directed line segments. Then in Anton et al. (2012, Sections 11.5 and 11.6) vectors are used to represent lines and planes and so, indirectly at least, vectors can be viewed as points. Anton makes a distinction between a vector with components $\langle a, b, c\rangle$ and the point (a, b, c) at the end of the arrow representing that vector if we place its tail at the origin. It is not clear whether this distinction helps students to unravel the roles of the different vectors in his vector representation of a line ($\mathbf{r} = \mathbf{r}_0 + t\mathbf{v}$) where the vectors **r** and $\mathbf{r}_0$ correspond to points on the line but the vector **v** corresponds to a direction parallel to the line. However, this is probably as close as Anton gets to thinking of a vector as matrix with just one column (or as an n-tuple).

On the other hand, Lay's introduction to linear algebra (Lay, 2012, Section 1.3) defines a vector to be a matrix with just one column (or row). Lay says that we 'can identify' such a vector with a geometric point, although this visualization 'is often aided by including an arrow' from the origin to this point. This allows Lay (2012, Section 1.5) to interpret $\mathbf{x} = \mathbf{p} + t\mathbf{v}$ as the equation of the line through **p** parallel to **v**. A physics view of vectors (but just as the set of all arrows or directed line segments) is mentioned in a footnote but not actually described until Section 4.1, when it serves briefly as an example of a vector space. Lay does not mention the dot product until he deals with the usual inner product in n-space (Chap. 6), while the cross product merits mention only in Chap. 8 where it features as a (presumably already known) trick for producing a normal vector in 3-space.

It is worth bearing in mind that students taking a beginning course in linear algebra will almost certainly meet both of these approaches (or minor variations of them) and that they may perhaps struggle to reconcile the two viewpoints. The books by Anton and Lay mentioned above suggest that mathematicians themselves solve this problem by ignoring whichever approach does not suit their purposes, and so we can hardly be surprised if students follow their example and keep these two worlds of vectors separate. I would like to suggest that, instead, there is much to be gained from discussing both viewpoints side by side.

The basic story is both an example of how mathematicians often work and an illustration of the power of abstraction. For our purposes, vectors can be seen as an invention of physicists, designed to solve problems about forces, velocities and so on, in the two or three dimensions that make up the real world around us. The mathematicians' role has been to abstract this notion (vectors are now just ordered pairs or triples) and to generalize it to higher dimensions. It doesn't even matter too much if this story is not strictly true historically, as it will almost certainly be true in terms of the student's own historical experience. Reconciling the different viewpoints takes a little extra effort (why does the physics definition of the dot product give the same result as the mathematical definition of the inner product?) but, as we saw above, current texts already do this without saying that this is what they are doing. Furthermore, by making this reconciliation of viewpoints explicit, we can engage the students in meta-level discussions about the nature of mathematics, an activity which some writers have suggested has the potential to deepen students' understanding of linear algebra (Hillel, 2000, p. 206).

Such diversions can seem quite time-consuming if you are teaching to a content-rich curriculum, but in this case you do not need to wait too long to illustrate the benefits of abstraction with a simple example. Indeed, as Fig. 11.1 shows, the same calculations that give us the component of a given force parallel to a given direction (Anton et al., 2012, Section 11.3) also give us a way to find the point on a given line which is closest to a given point (Lay, 2012, Section 6.2). Both interpretations of this calculation are a standard part of a linear algebra course (Anton gives the point interpretation of this calculation as an exercise at the end of Section 11.3, while Lay mentions the decomposing forces interpretation as an aside in Section 6.2) but by discussing the relationships between the interpretations we can open the students' eyes to how mathematics is done (Carlson, 1993, p. 45).

11.3 Why Do We Need All the Jargon?

There is no doubt that students meeting linear algebra for the first time are faced with a long sequence of new technical terms: linear combination, span, linear independence, subspace, basis and dimension, just to name the most basic ideas. Furthermore, we often use earlier terms in this sequence when defining the later terms, and so failing to understand one term can be an obstacle to understanding several others.

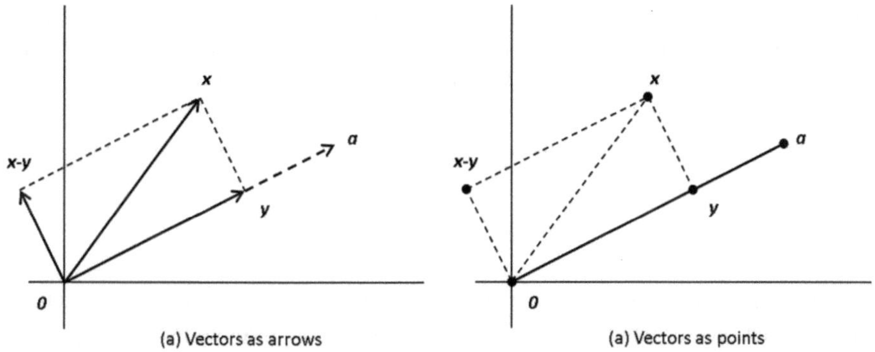

Fig. 11.1 Depending on the interpretation of vectors, the same calculation can yield either (**a**) the projection **y** of a vector **x** parallel to a vector **a**, or (**b**) the point **y** on the line through **0** and **a** which is closest to the point **x**

Many recent textbooks try to ease the strain on students by pacing the introduction of these terms, giving the students time to get used to the earlier, and perhaps simpler, new terminology. For example, Lay (2012) introduces linear combinations and span in Section 1.3, while discussing solutions to systems of linear equations. Linear independence is introduced in Section 1.7 (without much motivation, but with plenty of geometric interpretations) but subspaces are not introduced until Section 2.8 at a point in the exposition where the subspaces associated with a matrix (column, row and null spaces) can all be introduced. Poole (2011) follows a similar strategy, with linear combinations done early (Section 1.1) then span and linear independence (Section 2.3) and finally the other terms in the above list (Section 3.5).

Introducing the new terms gradually certainly gives the students more time to become familiar with the most basic ideas but, by itself, this strategy does not explain to the students why these ideas are important enough to warrant the invention of new names. This importance will perhaps become more evident when the ideas are used to solve problems or prove theorems, but relying on that delayed gratification runs the risk that students will have got lost in the fog (Carlson, 1993) long before they get the chance to experience some motivation. To paraphrase John Mason in an earlier chapter of this book, unless students appreciate where the concepts come from, linear algebra will remain a mystery.

Harel's three principles for learning linear algebra (Harel, 2000) offer us a useful framework when planning how to introduce this new terminology. The principle of necessity prompts us to let students experience for themselves the need to introduce a particular new term. Why might we need a new piece of terminology? It is tempting to say that we need it because it will be important later on but, if we are worried that delayed gratification is not a strong enough motivation, we may have to settle for more immediate needs, such as the desire to abbreviate a complicated or long-winded description. The principle of concreteness directs us to look for a concrete context, familiar to the students, in which the students can experience this need and from which they can abstract the new idea. Finally, the generalizability principle asks us to design the students' activities in such a way that students are able, and encouraged, to do this abstraction in a generalizable way.

If we look at recent textbooks, it seems that it is easy to follow this scheme for the idea of a linear combination. Algebraic expressions that we now call linear combinations

$$\boldsymbol{x} = s_1 \boldsymbol{v}_1 + s_2 \boldsymbol{v}_2 + \cdots + s_n \boldsymbol{v}_n$$

can arise naturally when we write down in vector form the solution to a system of linear equations, or when we describe points on a plane through the origin parallel to some given vectors. They also arise when we write down the general solution to a linear differential equation with constant coefficients, although in that case we need to broaden our notion of what a vector is. Introducing the term 'linear combination' simplifies the recitation of the above equation. Incidentally, this new term also focuses attention on the vector participants in the equation, at the expense of the scalars. This is probably a beneficial side effect if we envisage talking about whether these vectors form a basis, but it may be less desirable if we wish to interpret the above vector equation as a system of linear equations with the scalars as the unknowns.

The ideas of linear dependence and independence seem to be a bit more complicated and are perhaps more difficult to motivate. For Lay the primary notion is linear independence and he introduces the usual definition by connecting the equation

$$s_1 \boldsymbol{v}_1 + s_2 \boldsymbol{v}_2 + \cdots + s_n \boldsymbol{v}_n = 0$$

with his earlier discussion about the trivial solution to a homogeneous system of linear equations, where the vectors are the columns of the coefficient matrix and the scalars are the unknowns (Lay, 2012, Sections 1.7 and 1.5). On the other hand, Poole sees linear dependence as the primary concept and he defines it informally first, saying one of these vectors is a linear combination of the others, and then gives the usual definition, presenting it as equivalent to his informal definition but without having to commit to which particular vector it is a linear combination of the others (Poole, 2011, Section 2.3). These are fairly traditional approaches to linear independence but, away from the mainstream of linear algebra texts, researchers have tried more novel ways of following, consciously or not, Harel's three principles. In one interesting example, Wawro, Rasmussen, Zandieh, Sweeney, and Larson (2012) reported positive outcomes from a series of tasks which used a very concrete setting, journeys on a two-dimensional map using a choice of hover board or a magic carpet each of which can only travel parallel to a given vector. Other research teams have reported mixed results from attempts to build on students' experiences of geometric contexts (see Thomas et al., 2015 for an overview), citing in particular the dangers of generalizing from experiences in only two or three dimensions. I

What kind of geometric object do you usually get if you take all the scalar multiples of a given vector in 3-space? Are there any exceptions?
What kind of geometric object do you usually get if you take all the linear combinations of two given vectors in 3-space? Are there any exceptions?
What kind of geometric object do you usually get if you take all the linear combinations of three given vectors in 3-space? Are there any exceptions?

Fig. 11.2 Geometric questions leading to the idea of linear dependence

have had good responses from students by introducing linear dependence and independence through the following discussions and activities.

Firstly, we can see a need for, and then construct, the idea of linear dependence in two or three dimensions by discussing the question of what kind of geometric object we get if we look at all linear combinations of n vectors in 2-space, or in 3-space, where $n = 1, 2, 3$. See Fig. 11.2 for the questions you might ask in the three-dimensional case.

Setting the questions in the context of 2-space or 3-space means that the relevant objects (vectors, lines, planes and so on) can be represented by familiar physical props (such as pens, rods, sheets of paper or other flat surfaces). The focus of attention here is not so much deciding what usually happens, but rather trying to describe all the exceptions. For example, the scalar multiples of a single vector usually give us a line, except when that vector is zero. Similarly, students readily see that the linear combinations of two vectors in 3-space usually give a plane, but this time there are more exceptions. Clearly the zero vector will again cause problems, but the two vectors will also fail to generate a plane if the vectors are parallel. Again, it is relatively easy to see that the linear combinations of three vectors in 3-space usually give us all of 3-space, but this time there are even more exceptions. The exceptions that caused problems in the previous two cases will clearly cause problems again, but students may find it harder to see another exception, where one of the vectors lies in the plane of the other two. Initially, all these exceptions seem quite different, but a class discussion can (perhaps with a little guidance) home in on the fact that all the exceptions are captured by the single idea of one vector being a linear combination of the others. This idea will eventually get named 'linear dependence' but, as observed above, it is probably a good idea first for students to see this phenomenon arise in a setting which breaks out of the geometric confines of two or three dimensions.

One way to create such an experience is a similar exploration of homogeneous systems of linear equations, looking for how many solutions you usually get (exactly one, or infinitely many) when you solve m linear equations in n unknowns (see Fig. 11.3). This is rather less concrete than the magic carpet ride of Wawro et al. (2012) but it does have the advantages of both building on the students' experience of solving systems of linear equations, and preparing the ground for what Strang (1988, Section 2.4) calls the Fundamental Theorem of

How many solutions does a homogeneous system of three linear equations in three unknowns usually have? Are there any exceptions?
How many solutions does a homogeneous system of two linear equations in three unknowns usually have? Are there any exceptions?
How many solutions does a homogeneous system of four linear equations in three unknowns usually have? Are there any exceptions?
How many solutions does a homogeneous system of m linear equations in n unknowns usually have? Are there any exceptions?

Fig. 11.3 Questions which may lead to the idea of linear dependence among linear equations

Linear Algebra, or what Lay (2012, Section 4.6) and Poole (2011, Section 3.5) call the Rank Theorem.

This time the activity is conducted in a computer lab so that students can focus on the structure of the systems of equations and their solutions without getting bogged down in all the details of the actual solution process. To build on the students' intuitions about geometric objects, the discussion can begin with systems of equations in two or (as in Fig. 11.3) three unknowns, but once the discussion has been started there is no bound this time on the dimension of the 'spaces' being considered. The only constraint is the students' experience of solving 'large' systems of linear equations, or their willingness to try something even bigger.

The simplest case is one equation in one unknown.

$$ax = 0$$

This may seem too simple to be interesting, but starting here establishes an analogy with the previous discussion about geometric objects generated by linear combinations of vectors in 2- or 3-space. Thus the usual situation for this 'system' of equations is that there is exactly one solution, but there is an exception: if $a = 0$ then there are infinitely many solutions. So the zero equation ($0x = 0$) is the exception here.

The next simplest case is a homogeneous system of two linear equations in two unknowns

$$\begin{aligned} a_{11}x_1 + a_{12}x_2 &= 0 \\ a_{21}x_1 + a_{22}x_2 &= 0 \end{aligned}$$

and this may be best explored by using the familiar geometric interpretation of these equations. Geometrically, each equation represents a line through the origin. Usually these lines are different and they meet at a single point (the origin). What are the exceptions? As in the case of the single equation, there is an exception if one of the equations is the zero equation. But there is another exception, where the two lines are the same, and this corresponds to the case where one equation is a (constant or scalar) multiple of the other equation.

The situation with systems of three equations offers yet another type of exception. If there are three unknowns then the system usually has exactly one solution but, in addition to the exceptions we have already seen, there is now the possibility that one equation is a linear combination of the other two. Typically this will be diagnosed by the fact that two elimination steps are needed, subtracting multiples of two of the equations from the third one, and that this results in that third equation becoming the zero equation. The analogy with the previous scenario (looking at linear combinations of vectors in two or three dimensions) should be clear by now.

To understand the situation for larger systems, the students may need to review what happens when you carry out Gaussian elimination to solve such systems. This algorithm can be viewed either in terms of elimination operations on the actual equations, or else in terms of elementary row operations on the corresponding matrix of coefficients. From the equation viewpoint, the key elimination operation is to subtract a multiple of one equation from another equation, thus getting a new and simpler equation:

$$(equation)_i \to (equation)_i - s(equation)_j$$

Now, the symptom of all the exceptional cases is that this process eventually yields the zero equation. In the matrix viewpoint, this means that the row echelon form of the coefficient matrix contains a row of zeros. It's not too hard to see that this happens because one of the equations is a linear combination of the other equations. However, this is probably easier to do verbally than as a written argument, where the notational complications might be a bit off-putting.

These explorations thus reveal a similar pattern to the one observed in the case of linear combinations of vectors in two or three dimensions. This time, however, the linear dependence relationship applies to the equations, or to the rows of the coefficient matrix, with the equations or rows being viewed as vectors. Notice that in both cases linear dependence arises in the form of one vector being a linear combination of the others. I suspect that this is an easier way to approach the idea than via the traditional definition in terms of the more symmetrical relationship

$$s_1\, \boldsymbol{v}_1 + s_2\, \boldsymbol{v}_2 + \cdots + s_n\, \boldsymbol{v}_n = 0.$$

Poole (2011, Section 2.3) seems to agree on this point, as he mentions dependence relations like $\boldsymbol{w} = 3\boldsymbol{u} + 2\boldsymbol{v}$ before using the symmetrical version in his official definition, and then shows that a set of vectors is linearly dependent exactly when one of them is a linear combination of the others.

The main point, though, is that the need for a definition comes from looking at some examples. The definition doesn't just appear out of nowhere, already fully formed like Athena emerging from the forehead of Zeus.

11.4 More About Experiments and Report Writing

The idea of students using software like MATLAB to explore and discuss mathematical ideas has been around for a while now. Carslon, in his commentary on the recommendations of the Linear Algebra Curriculum Study Group (Carlson, 1993, p. 45), noted that 'students learn best, as we do, by active involvement—solving problems, making conjectures and communicating with others'. Such activities were also central to reform calculus projects such as the Connected Curriculum Project (1997–2003) where students were set 'thought-provoking questions that require written answers, [and] summary questions that enable students to see the forest as well as the trees'.

An unplanned benefit of the above experimental approach to linear dependence was the discussions which arose among students when they met an unexpected cognitive conflict. They knew from earlier work on solving systems of linear equations via Gaussian elimination that there are simple examples of, say, three equations in three unknowns which have no solutions, or which have infinitely many solutions. But, in the above experiments, computer simulations using MATLAB and its random matrix generator only ever gave examples of three equations in three unknowns which had exactly one solution. Unexpected behaviour like this is a great conversation starter. After making sure their neighbour is getting the same strange behaviour, students look for explanations. A surprisingly common reaction was to question the randomness of MATLAB's random matrix generator.

The true explanation of this phenomenon can be found by looking at systems of linear equations from an angle that is probably unfamiliar to most students. So maybe it deepens the students' understanding. The underlying problem in the linear dependence experiments is that, just as the diagonal of a square has zero area, so the set of exceptions for systems of three equations in three unknowns (see Fig. 11.3) has zero measure in the set of all such systems. So the chances of MATLAB conjuring up an exceptional matrix are very, very small. One determined student reported doing 100,000 trials without succeeding in finding an exception! Other students struggled with the idea that an event can have probability zero but still be possible.

Such experiments illustrate the fact that even if something never happens in repeated computer simulations it could still be possible. If you want to be sure that something never (or always) happens, then you need a logical proof. Experiences like this seem to offer an opportunity for the students to experience a need, in line with Harel's principle of necessity, for logical proofs and counter examples.

As we have just seen, some experiments always give the same result because, although exceptions exist, they are too rare to show up in a relatively brief lab session. For example, almost all homogeneous systems of three linear equations in three unknowns have exactly one solution, and so that is all that we see in the experiments. We may need to call on other resources, such as a geometric intuition about the intersection of three planes in 3-space, in order to determine that there are exceptions to the experimentally observed pattern. On the other hand, there may be occasions where the other resource needed is a logical proof that the observed behaviour is the only possible behaviour. This happens (to take a simple example) when we solve a homogeneous system of three linear equations in four unknowns. Exceptional cases in linear algebra sometimes correspond to a set of measure zero (for example, when a determinant is zero rather than nonzero) and so such situations will be resolved not by computer simulations but by a mathematical search for a proof or a counterexample. I have not yet explored this idea in any systematic way, but maybe students can be brought to an appreciation of the need for proofs by a series of similar experiments.

11.5 Conclusion

I am conscious of the fact that, in a book mostly about high school algebra, I have been discussing a topic which is not covered (in my country, at least) until university level mathematics courses. However, although the gap can seem huge in terms of the level of mathematical content, there is less of a gap in terms of pedagogical approaches. Just as John Mason insists in an earlier chapter that learners need to appreciate how equations emerge from an expression of generality if they are to understand the role that algebra can play in solving those equations, so I am suggesting that students will benefit from seeing the abstraction of linear algebra emerge from experiments or problem solving rather than from a predetermined set of axioms. This means that linear algebra needs to evolve from the students' experiences, rather than appearing out of nowhere, already fully formed.[1]

In Sect. 11.2, we saw how this evolution can be based on students' prior experiences. Thus we can acknowledge the physicists' view, which most students have already met, of vectors as quantities having both magnitude and direction, with representations as arrows or directed line segments, and build connections between these ideas and the more abstract and more general (mathematical) idea of a vector as an ordered n-tuple of real numbers. By connecting these two viewpoints, students can see how the same calculation (in the example in Sect. 11.2 this was the calculation of the projection of one vector in the direction of another vector) can simultaneously solve two quite different-looking problems. So the effort of abstracting can be justified from a cost-benefit point of view.

In Sect. 11.3, I suggested that some of the concepts of linear algebra could also be allowed to emerge from carefully designed discussions or experiments experienced by the students in their first linear algebra course. This need not always mean cramming even more material into an already full course. In some cases it is simply

[1] The same idea occurs in more general senses in other chapters of this book. For example, Mercedes McGowen discusses how prior knowledge can support or impede new learning.

a question of delaying a definition until the students can see a need for it (Harel, 2000). The experiences that may prompt such a feeling are often already in current textbooks but placed after the definitions, as if mathematicians had actually thought of the definition first. In my example, we saw linear dependence emerging from a discussion of *exceptions* to the pattern that the linear combinations of one, two or three vectors in 3-space produce lines, planes or all of 3-space (respectively). But sometimes extra room may need to be made in order to develop richer experiences for the students. For example, I suggested that MATLAB experiments on homogeneous systems of linear equations may help develop a deeper feeling for linear dependence which breaks out from the confines of two or three dimensions used in the previous example. Such experiments also have the advantage, as we saw in Sect. 11.4, that they can naturally inspire a need for logical proof (or counter-examples) as a way to distinguish between situations which look identical to a computer, namely, phenomena which occur almost all the time and those which occur always.

The general idea, both in this chapter and in Mason's, is to give students a context in which the importance of ideas, and the need to name them, will become apparent in the natural course of events. By following this approach we are often imitating the historical development of many of the abstract ideas we use in modern mathematics. Thus we are not only teaching our students the expected new content (in my case, linear algebra, but in Mason's case, symbolic algebra) but we are also showing students how mathematics is actually done by practising mathematicians.

References

Anton, H., Bivens, I., & Davis, S. (2012). *Calculus: Early transcendentals*. Hoboken, NJ: Wiley.

Carlson, D. (1993). Teaching linear algebra: Must the fog always roll in? *The College Math Journal, 24*, 29–40.

Carlson, D., Johnson, C. R., Lay, D. C., & Porter, A. D. (1993). The linear algebra curriculum study group recommendations for the first course in linear algebra. *The College Math Journal, 24*, 41–46.

Connected Curriculum Project. (1997–2003). Retrieved from https://www.math.duke.edu//education/ccp/resources/teach/index.html

Dorier, J.-L. (Ed.). (2000). *On the teaching of linear algebra*. Boston: Kluwer.

Harel, G. (2000). Three principles of learning and teaching mathematics. In J.-L. Dorier (Ed.), *On the teaching of linear algebra* (pp. 177–189). Boston: Kluwer.

Hillel, J. (2000). Modes of description and the problem of representation in linear algebra. In J.-L. Dorier (Ed.), *On the teaching of linear algebra* (pp. 191–207). Boston: Kluwer.

Khoury, J. (2006). *Applications of linear algebra*. Retrieved October 29, 2015, from http://aix1.uottawa.ca/~jkhoury/app.htm

Lay, D. C. (2012). *Linear algebra and its applications* (4th ed.). Reading, MA: Addison-Wesley.

Poole, D. (2011). *Linear algebra: A modern introduction* (3rd ed.). Boston: Brooks/Cole.

Strang, G. (1988). *Linear algebra and its applications* (3rd ed.). San Diego: Harcourt Brace Jovanovich.

Thomas, M. O. J., deFreitas Druck, I., Huillet, D., Ju, M.-K., Nardi, E., Rasmussen, C., et al. (2015). Mathematical concepts in the transition from secondary school to university. In

S. J. Cho (Ed.), *The Proceedings of the 12th International Congress on Mathematical Education*. New York: Springer.

Wawro, M., Rasmussen, C., Zandieh, M., Sweeney, G., & Larson, C. (2012). An inquiry-oriented approach to span and linear independence: The case of the Magic Carpet Ride sequence. *PRIMUS, 22*(8), 577–599.

Chapter 12
School Algebra to Linear Algebra: Advancing Through the Worlds of Mathematical Thinking

Sepideh Stewart

Abstract Linear algebra is a core subject for mathematics students and is required for many STEM majors. Research reveals that many students struggle grasping the more theoretical aspects of linear algebra which are unavoidable features of the course. Working with vectors and understanding new concepts through definitions, theorems, and proofs all indicate that a sudden shift has occurred, and despite carrying the name "algebra," in many respects linear algebra is significantly more complex than school algebra. In this chapter we will employ the Framework of Advanced Mathematical Thinking (FAMT) to describe the type of thinking that is required for linear algebra students to succeed at college level.

Keywords Advanced mathematical thinking • Linear algebra • Three worlds of mathematical thinking • Algebra • APOS

12.1 Introduction

Many students find the sudden shift from high school to linear algebra difficult. In an interview a group of linear algebra students were asked: Did you notice much similarities or differences between high school algebra and linear algebra? Their responses were:

- *In high school we started with examples, we didn't really touch on proofs. It has been the biggest difference here. It's probably why I'm struggling with it.*
- *In the beginning, yes, I thought it was identical to what I learned in high school, was just easy and then it went a 180 degrees from simple high school stuff to thinking of what's the proofs and everything.*
- *In high school algebra obviously just the rote computation of things, it is very difficult to convey a understanding of what this is all leading to so that can probably be a lot more tedious and it is just like mechanical, you just need to*

S. Stewart (✉)
Department of Mathematics, University of Oklahoma, 601 Elm Ave, Norman, OK 73019, USA
e-mail: sepidehstewart@ou.edu

S. Stewart (ed.), *And the Rest is Just Algebra*, DOI 10.1007/978-3-319-45053-7_12

learn these rules and apply them whereas in linear algebra it is certainly much more exciting the further you go.

One student referred to the computation part of the course, for example, finding a basis for the *kernel* and *Range* of a matrix, as "*the math part.*"

Most linear algebra textbooks start by introducing the systems of linear equations and move into the study of matrices. As the course progresses, the concepts carry more theorems and with them come challenges for many students. Carlson (1997) noticed his students were able to solve systems and handle the matrix multiplications easily. However, he was concerned that: "*when we get to subspaces, spanning, and linear independence, my students become confused and disoriented. It is as if a heavy fog has rolled in over them, and they cannot see where they are or where they are going. And I, as a teacher, become disheartened, and question my choice of profession*" (p. 39).

The research on linear algebra over the past decade has revealed the nature of students' difficulties and thought processes (e.g., Briton & Henderson, 2009; Hannah, Stewart, & Thomas, 2013, in press; Stewart & Thomas, 2009; Wawro, Sweeney, & Rabin, 2011; Wawro, Zandieh, Sweeney, Larson, & Rasmussen, 2011). Despite the fact that we now have more evidence that these problems in fact exist, "*... still the fog rolls in, and students feel as though they have been taken to a new world*" (Briton & Henderson, 2009, p. 963). In a linear algebra study (Stewart & Thomas, 2009) one student made it clear that he did not know how to explain the definitions of the given concepts, but claimed he could calculate (see Fig. 12.1). Apparently, this is not a rare occurrence, as Day (1997) confirmed, her engineering and scientific colleagues remembered very little about their undergraduate linear algebra courses. Those courses had covered little about properties of matrices, and apparently the abstract concepts that were covered did not sink in. "*These colleagues could not state sensible definitions of concepts like linear independence and span, and their geometric understanding of such concepts was nil*" (p. 71).

Dorier (1990) is concerned that teachers are emphasizing "*less and less on the most formal part of the teaching (especially at the beginning) and most of the evaluation deals with the algorithmic tasks connected with the reduction of*

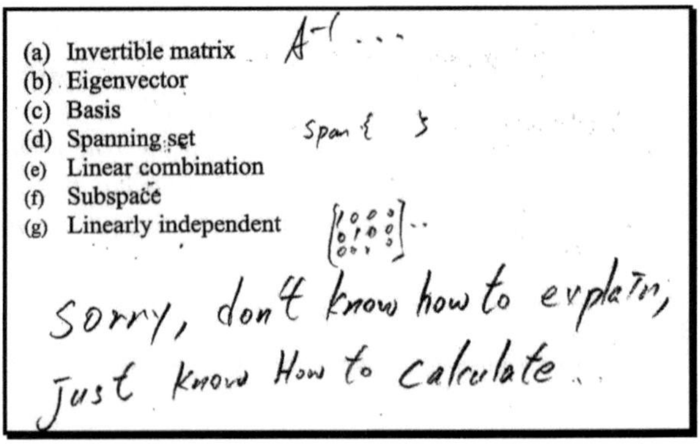

Fig. 12.1 An honest linear algebra student's response

matrices of linear operators"(p. 28). Carlson (1997, p. 40) agrees and adds that: "*These are concepts, not computational algorithms like Gaussian elimination and matrix multiplication.*" According to Sierpinska, Nnadozie, and Okta (2002, p. 2) this is a "*waste of students' intellectual possibilities.*" In their views "*linear algebra, with its axiomatic definitions of vector space and linear transformation, is a highly theoretical knowledge, and its learning cannot be reduced to practicing and mastering a set of computational procedures*" (p. 1). Andre Revuz who wrote the Preface for the book: *On the Teaching of Linear Algebra* (Dorier, 2000, p. xv) reminds us of an important reality that:

> A common preconception among mathematician is that in order to teach mathematics well, all that is necessary is to know the subject well. The teaching of linear algebra provides a striking counter example. The theory is well developed, those who teach it know it personally well . . .yet the students do not understand.

12.2 Framework of Advanced Mathematical Thinking (FAMT)

Over the last decade, we have employed Tall's (2004, 2008, 2010, 2013) framework of embodied, symbolic, and formal mathematical thinking along with Dubinsky's (Dubinsky & McDonald, 2001) Action, Process, Object, and Schema (APOS) theory and built a framework (Stewart & Thomas, 2009), namely the Framework of Advanced Mathematical Thinking (FAMT). This framework (see Fig. 12.2) has enabled us to investigate students' conceptual understanding of major linear algebra concepts (Hannah, Stewart, & Thomas, 2013, 2014, 2015, 2016; Stewart & Thomas, 2009, 2010; Thomas & Stewart, 2011). The natural blend of these two learning theories provides an ideal platform to analyze students' thinking in the context of main concepts in linear algebra, for example, vectors, linear combinations, linear independence, basis, span and eigenvalues and eigenvectors.

Tall (2010) defines the worlds as follows: The *embodied world* is based on "*our operation as biological creatures, with gestures that convey meaning, perception of objects that recognise properties and patterns...and other forms of figures and diagrams*" (p. 22). Embodiment can also be perceived as giving body to an abstract idea. The *symbolic world* is based on practicing sequences of actions which can be achieved effortlessly and accurately as operations that can be expressed as manipulable symbols. The *formal world* is based on "*lists of axioms expressed formally through sequences of theorems proved deductively with the intention of building a coherent formal knowledge structure*" (p. 22).

Dubinsky and McDonald (2001) define action, which is somewhat external and requires either explicit or from memory, step-by-step instructions and rules on how to perform a certain task. Once an action is repeated and it is reflected upon by the individual, it may be interiorized into a process. The individual can successfully think of a process as an object, when he or she is able to "*reflect on operations applied to a particular process, becomes aware of the process as a totality, realizes*

Worlds APOS	Embodied World	Symbolic World Algebraic Rep.	Symbolic World Matrix Rep.	Formal World
Action	Can apply a specific transformation and multiply it by a specific scalar. 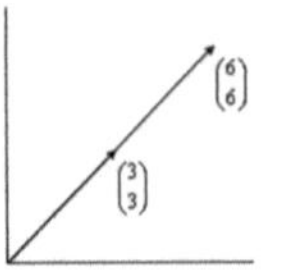	Can simplify $A\mathbf{x} = \lambda\mathbf{x}$ to $(A - \lambda I)\mathbf{x} = 0$ in an specific example to find the eigenvectors	Can find the eigenvalues and eigenvectors of the matrix $\begin{bmatrix} 1 & 2 \\ 3 & 4 \end{bmatrix}$ Can find a basis for the eigenspace	
Process	Can apply a transformation and show it for a general case. 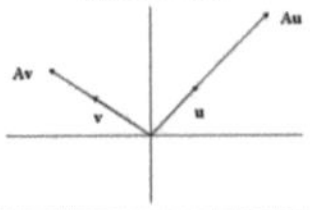	Can see that there are infinitely many eigenvectors associated with an eigenvalue Understand the process $(A - \lambda I)\mathbf{x} = 0$ (realises that $Ix = x$)	Can see that every scalar multiple of an eigenvector is also an eigenvector	Can relate to a basis for the eigenspace An eigenspace consists of the zero vector and all the linearly independent eigenvectors
Object	Vectors that are transformed by a matrix, and stretched or shrunk in the same direction by a scalar	Understands that for $A\mathbf{x} = \lambda\mathbf{x}$ in each side of the equation the final object is a vector		Understanding the definition of eigenvectors and eigenvalues

Fig. 12.2 FAMT: Framework of Advanced Mathematical Thinking

that transformations can act on it, and is able to actually construct such transformations. In this case, the process has been encapsulated to an object" (Asiala et al., 1996, p. 11).

In this chapter we employ FAMT to describe some of the complexities of understanding linear algebra for the first year students and the types of thinking that are required at this level.

12.3 Connecting the Core Linear Algebra Concepts

The ability to connect core ideas in linear algebra and indeed in any advanced mathematics topics is one of the most important proficiencies that we ought to convey to our students. Failing this objective creates major issues for many students (a process-formal view) and impedes their growth and appreciation of concepts. Here are some possible tasks to foster this type of thinking in the classroom.

12.3.1 Concept Maps

One way to foster students' abilities in connecting the main ideas is by asking them to draw a concept map of the core concepts. In a study by Stewart (2008) concept maps were used as a tool to detect whether students could relate the main concepts to each other. The most logical map which showed the progression of concepts and how they were formed was drawn by a Ph.D. graduate (see Fig. 12.3a) as he placed linear combination at the center, as he acknowledged that many concepts are built

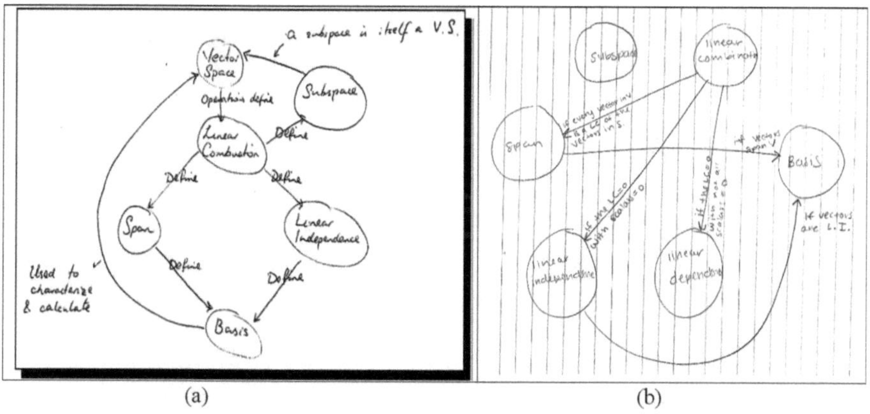

Fig. 12.3 An illustration of a concept map by a Ph.D. graduate (**a**) and a linear algebra student (**b**)

based on this concept. This, however, was not the case with the majority of students, who were not thinking about the relationships in which they were linking their concepts. It was also noted that although some students displayed an understanding of a concept in the test and interview, they did not display them in their concept maps. Although the Ph.D. graduate had no practice with drawing concept maps (as this is not a common practice in mathematics courses), he was able to produce a clear map illustrating his thoughts.

In another study by Hannah et al. (2015), students' concept maps gave some insight into how students were thinking about the concepts and connecting the ideas together, with some drawing more well thought-out maps than others. Interestingly, a number of students were unsure about where they should place the concept of subspace and a few did not connect this concept with any other concepts and left it on its own (see Fig. 12.3b).

As part of a recent class activity, while reflecting on the chapter on vector spaces, one linear algebra student wrote:

> *Chapter 4* [The vector spaces] *was like a very juicy peach on a high branch just outside of my reach. I understood the concepts as they were explained, but the entirety of the connections to each other and previous concepts remained elusive. The concept map activity made me further realize how jumbled my brain is with regards to what we've learned.*

Meel's (2005) investigation of the use of concept mapping in linear algebra suggests that we cannot draw strong conclusions about students' understanding from this method since it may not be "reliable." He suggests that "*concept mapping should best be used as an instructional tool rather than relied on as an assessment tool*" (p. 7). Although, we do not depend on concept maps for assessment, it is a useful tool in inspecting students' thought processes. As Duval (2006, p. 104) asserts: "*research about the learning of mathematics and its difficulties must be based on what students do really by themselves, on their productions, on their own voices.*" Concept maps are ideal tasks as part of homework or class activity.

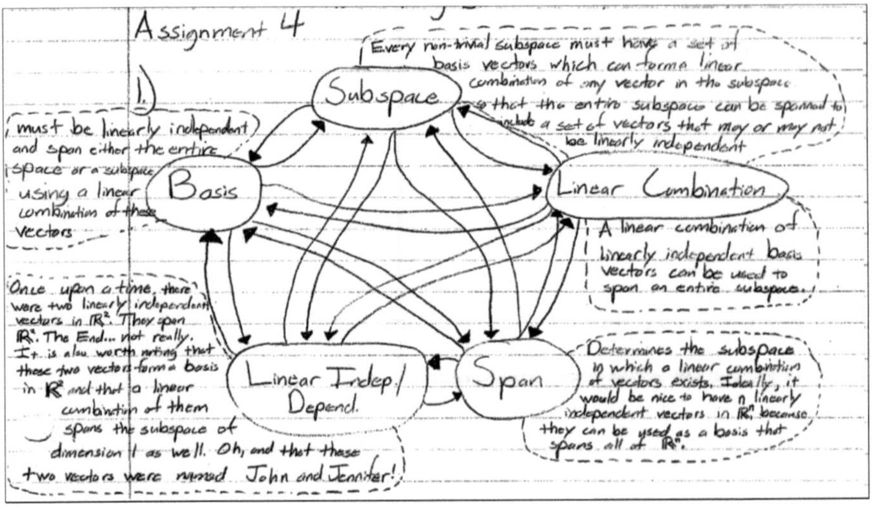

Fig. 12.4 Student's concept map

In author's experience, students are not pressured to look for answers elsewhere and usually spend a considerable amount of time articulating their drawings (see Fig. 12.4).

12.3.2 Journal Writing: Linking the Concepts for a Given Set of Vectors

Writing and describing terms may not be as common practice in a linear algebra classroom, but can help students to articulate their thoughts. In Duval's (2006) notion, it will allow students to express their voices. It also helps the instructor to detect early signs of misconceptions. In a study by Hannah et al. (2016), the following question was posed to a group of 162 linear algebra students at a term test:

Question. Consider the vectors $\mathbf{u} = [1, 0, 0]$, $\mathbf{v} = [0, 2, 0]$, $\mathbf{w} = [3, 4, 0]$. Write a short paragraph about **u**, **v**, and **w**. Your paragraph should be at most 75 words long, but should include as many as possible of the following technical terms from Linear Algebra: *basis, dimension, dependence relation, linear combination, linearly dependent, linearly independent, span, subspace*.

The results showed that most students found this question difficult (average score was 3.8/6). Only one student drew a diagram illustrating **w** as a linear combination of **u** and **v**, indicating that no embodied or geometric thinking was necessary. While most students worked in the formal world, only seven students set up a matrix for row reduction, indicating that the symbolic world actions and processes were not necessary or as helpful (see Fig. 12.5).

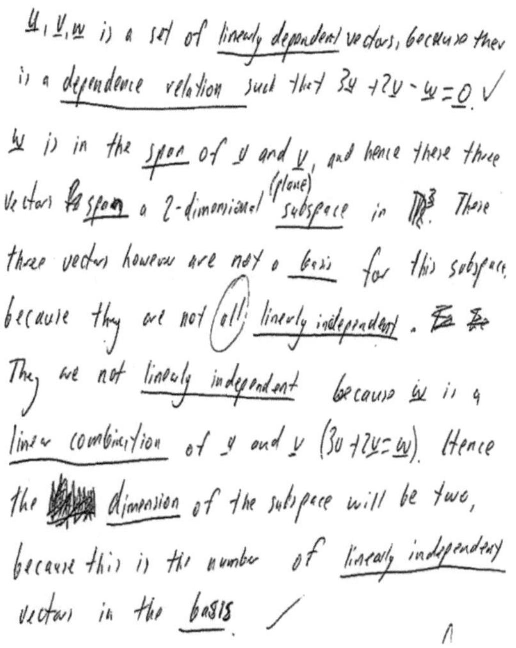
u, v, w is a set of linearly dependent vectors, because there is a dependence relation such that 3u + 2v − w = 0 ✓
w is in the span of u and v, and hence these three vectors span a 2-dimensional (plane) subspace in ℝ³. These three vectors however are not a basis for this subspace because they are not (all) linearly independent.
They are not linearly independent because w is a linear combination of u and v (3u + 2v = w). Hence the dimension of the subspace will be two, because this is the number of linearly independent vectors in the basis ✓

Fig. 12.5 Student's response to the test question

12.3.3 Need for Process Level Thinking: A Shift from Symbolic Manipulations of Matrices

The next set of examples are designed to encourage a process-formal level thinking without a need for calculation using matrices. Needless to say that students must know how to find the determinant of a matrix. Our main goal in the next set of examples (see Fig. 12.6) is to train students to use the given information and relate it to the solutions of the system or inverse of the matrix, etc. Early engagement in these activities sets the scene for moving away from a need for just calculations. Whole class discussions around relating the core ideas help the transition to this type of thinking.

Some students have misconceptions regarding the above questions which endure to the end of the course. For example, in response to Q1 in a final examination, students wrote: "*no solution*," "*yes, det(A)* = *0 therefore row equiv. to* I_n," "*1 trivial solution*," "$Ax = 0$ *has a trivial solution, and the system* $Ax = b$ *has a unique solution*." The following (see Fig. 12.7) are sample students' responses to Q2.

Q1.

[**4 marks**] Let A be a $n \times n$ matrix with det(A)= 0.

(i) Does A^{-1} exist? Justify your answer.

(ii) How many solutions does the homogeneous system $A\mathbf{x} = \mathbf{0}$ have?

Q2.

(c) [**3 marks**] Let A be an $n \times n$ matrix such that det (A)= 0. Write three more statements that are equivalent to this statement.

Fig. 12.6 Final examination and test questions

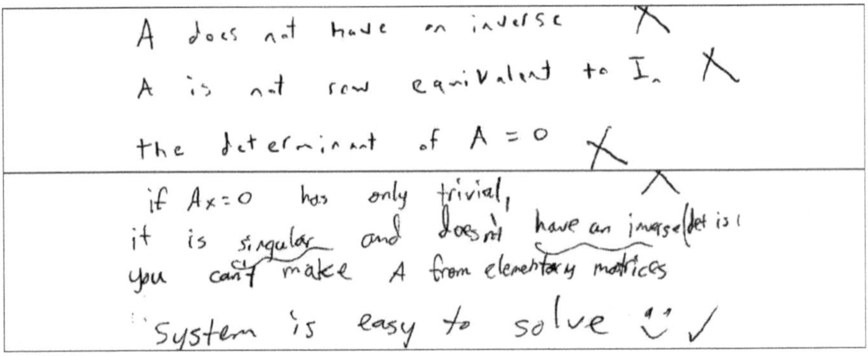

Fig. 12.7 Students' responses in a test

12.4 Representation of Concepts and the Ability to Transfer Between the Three Worlds

The ability to switch from one setting to another, for example, being able to see the linear equation $a_{11}x_1 + a_{12}x_2 + \cdots + a_{1n}x_n = 0$ as the n-tuple $(a_{11}, a_{12}, \ldots, a_{1n})$, i.e., from the linear combination of unit vectors to the rank of a set (Dorier, 1990; Robert, 1985, cited in Dorier, Robert, Robinet, & Rogalski, 1994) gives the students the opportunity to see the same problem from a different angle and, as a result, they have more control over solving the problem.

12.4.1 The Subtlety of Algebraic Representation of Vectors

Questions involving algebraic representation of arbitrary vectors are more complex to unravel than finding whether a set of given vectors in R^4 are linearly independent. The following question (see Fig. 12.8) was designed (Hannah et al., 2016) to examine a raft of students' abilities in moving between the worlds (embodied,

(a) Suppose that **u**, **v**, and **w** are nonzero vectors in $\mathbb{R}^3$ such that $\mathbf{u} = 2\mathbf{v} + 3\mathbf{w}$
   i. Draw a diagram to illustrate the relationship between **u**, **v**, and **w**.
   ii. Use the appropriate technical terms from linear algebra to describe the relationship between **u**, **v**, and **w**.

(b) Decide whether the vectors [0 2 0 1], [1 3 0 0], [0 4 1 0] are linearly independent.

(c) Suppose that **u**, **v** are linearly independent vectors in $\mathbb{R}^3$.
   i. Give a geometric description of the span of **u** and **v**.
   ii. Which of the following sets of vectors could be a basis for $\mathbb{R}^3$?
      (α) **u**, **v**, **u** + 2**v**.
      (β) **u**, **v**, **u**×**v**.
      (γ) **u**, **v**, **u** + 2**v**, **u**×**v**.

Fig. 12.8 Question from a term test

symbolic, and formal) as well as their ability to think in the action, process, and object level.

Due to the multilayered nature of this question, students did not perform well. The mean score for this whole question was 6.55 out of 12. For example, only 67 % were able to draw a diagram illustrating an embodied-process level thinking for part (a)i.

12.4.2 The Power of Visualization and the Ability to Transfer Between the Worlds

Although pictures are very useful, they are not always in forefront of students' thinking (Stewart & Thomas, 2009). In a class of 28 students no one used a diagram as part of showing the following proof (see Fig. 12.9), although a diagram was shown in class as part of the algebraic proof. Students instinctively think that a picture will not be sufficient and professors may not accept it as proofs, therefore it would not be necessary to draw one.

Although many students are confident to calculate the eigenvalues and eigenvectors, moving between the worlds causes considerable difficulties. The idea behind these concepts is to study equations in the form of $Ax = \lambda x$ and look for vectors that are transformed by A into a scalar multiple of themselves. Although the symbolic manipulations are necessary, in order to have an overall intuition for the concept it is useful for the learner to be exposed to the geometrical side of the concept. As Keith (2001, p. 156) describes in her book "Visualizing Linear Algebra with Maple" eigenvalues and eigenvectors "could be said to be the culmination of a linear algebra course, yet students frequently do not have a good understanding of them."

In the following question (see Fig. 12.10) students had to recognize geometrically that each of the three vectors satisfied the eigenvector definition, link this to the matrix size, and see a contradiction. A 2×2 matrix cannot have three independent vectors all with this property since it can only have at most two eigenvalues.

(c) [**4 marks**] If **u** and **v** are vectors in R^2 or R^3, prove that:

$$\mathbf{u} + \mathbf{v} = \mathbf{v} + \mathbf{u}$$

Fig. 12.9 Test question

If A is a 2 by 2 matrix, explain why the picture below is not possible.

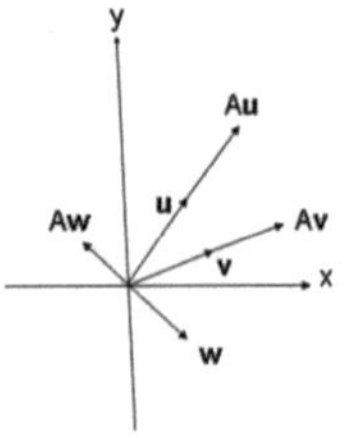

Fig. 12.10 A nonroutine question on eigenvalues and eigenvectors

The ability to recognize the eigenvectors in a different representation rather than the usual symbolic one does not occur naturally and often causes difficulties for many students.

Of the 42 students in a case study (Stewart, 2008), 14 did not respond, and only 6 were able to justify why such diagram did not make sense in this case. Some of their explanations are given below:

> *"The picture above implies A has 3 eigenvectors of different directions (but if A is 2 × 2, it has a maximum of 2 eigenvectors of different directions)." "If A is 2 × 2 matrix, it can have maximum of 2 linearly independent vectors in its basis. Therefore one of Aw, **Au** and **Av** must be impossible." "Diagram shows 3 eigenvalues/eigenvectors a 2 × 2 matrix should have only 2."*

Others were unable to relate the picture to the concept of eigenvector, and instead related it to the basis for a space. They wrote comments such as "*since there are only 3 vectors it will generate a space,*" "*because you don't need that many vectors to span the plane,*" and "*maybe too many dimensions.*"

As illustrated in Fig. 12.11a, one student performed many irrelevant calculations in response. In the interview, when the researcher pointed to his calculations, he said: "*Yeah, I didn't understand, I was trying to understand... because until now I don't understand what that means, I can't understand. Those vectors look very easy but I can't solve it.*"

So the researcher said: "*So in this situation, where you have no ideas, what is your best technique?*".

He replied: "*From the basic, just like I did here, I forgot about those eigenvalues and eigenvectors I started from the basic of basic.*"

The researcher then asked: "*Do you try to use numbers or symbols or do you mainly think about the theory, where the questions come from?*".

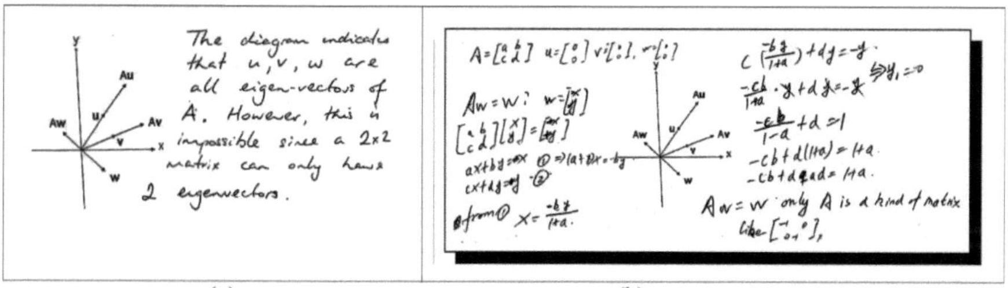

(a) (b)

Fig. 12.11 A Ph.D. graduate (**a**) and a student's (**b**) responses to the question on eigenvalues and eigenvectors

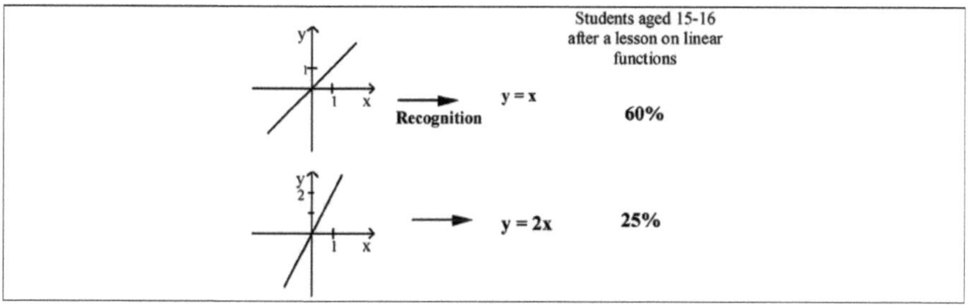

Fig. 12.12 Difficulties going from one register to another

He replied: "*Actually I use symbols, numbers are special symbols, and definition, I can't remember definition.*"

In this case a student who cannot remember the definition (formal ideas) and has probably never seen the eigenvectors geometrically (embodied representations) has no choice other than to revert to symbolic manipulation (actions) of what he calls "basic ideas" .

Duval (2006) noted that to construct a graph (see Fig. 12.12), most students have no difficulties as they follow a certain rule, "but one has only reverse the direction of the change of register to see this rule ceases to be operational and sufficient" (p. 113).

12.4.3 Definitions, Theorems, and Proofs

The symbolic representations of concepts which are rooted deeply in the formal world generate powerful definitions which play an important role in understanding the linear algebra concepts.

The ability to apply the definitions to various problem-solving situations is essential. In a study by Stewart (2008) in response to define the term linear combination, 45.5 % of students did not write any answer, and the remaining

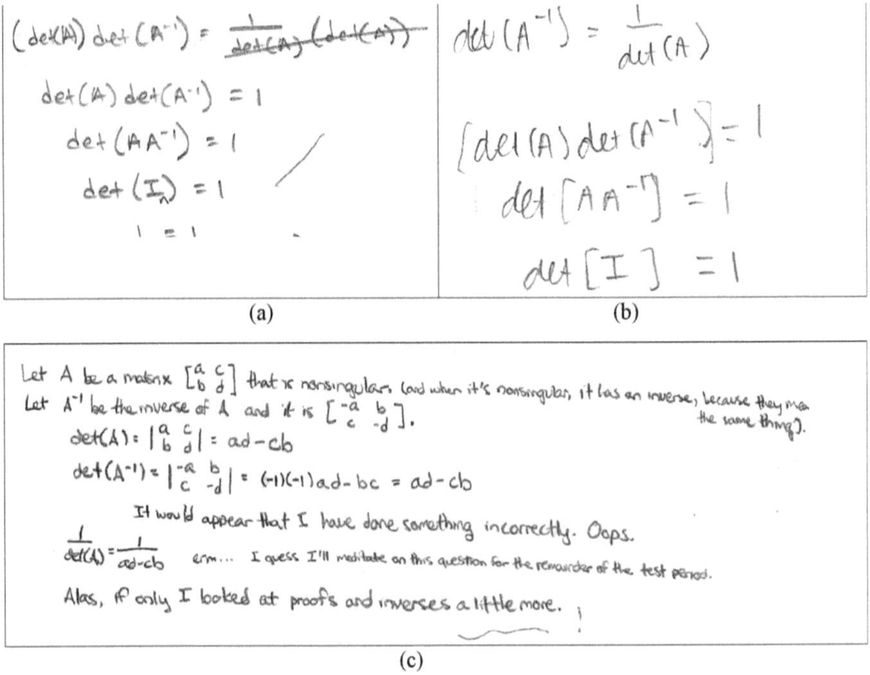

Fig. 12.13 Students' proofs

54.5 % only gave procedural or incomplete responses. For example, a student said: "something like $xv + yu$ $x,\ y \in R$." In an interview the same student said "linear combination, hmm … I can't quite remember the definition, I can just remember those forms something like $b = x_1v_1 + x_2v_2$ and something like that and x belong to R. I only can remember these things." When he was asked for further explanation he said: "Hmm …difficult! Linear combination is an object class in a space formed by the two vectors and x, y are scalars, this is my understanding of linear combination." This clearly demonstrates his lack of knowing the definition and not having an object view of the concept in general.

As for theorems and their proofs, most linear algebra students do not have a complete appreciation of their roles. The traditional methods of instructions, namely, writing the proofs on the board and students copying and later memorizing them, has not been fruitful. In a study by Hannah et al. (2014), when a group of linear algebra students were asked to name which of the three worlds (embodied, symbolic, or formal) they felt most comfortable, eight of the ten students selected the symbolic world. One student added that: "*Symbolic is the easiest for me but I enjoy formal thinking the most*" (p. 246).

In a study examining students' proving skills (see Fig. 12.13), we noted how students (a) and (b) began their proofs by assuming what they were trying to prove.

Also, student (c) tried to resort to the symbolic-matrix world and made an attempt in calculating the determinant of an arbitrary matrix.

As part of a larger study, students' opinion regarding proofs was asked. Although students had differing opinions and the type that they preferred, they mostly agreed that the proofs were necessary.

- *I like knowing the proof because, yeah it's nice to have a definition or just an example but knowing why it works, it helps you solve all those steps in between that you solve in the proof to get to your final theorem.*
- *Prove that what you're doing is not just fallacy or it's not just something that somebody made up. It's real. It's a law.*
- *Ideally the purpose of a proof is to show why a theorem works starting from beginning to end and ensure that it works for all cases. To me it exists to be a pain in the butt.*
- *I like proofs that use the math and not the words.*
- *I just like proofs that are not necessarily, one or two line proofs I always find they are more difficult because like I said they're either a restatement of a theorem in a weird way or they're just dumb and I don't understand the point of them but like slightly longer than that where there is stuff that you do with them. Maybe if they involve a little bit of math too. Ones that definitely just involve words I hate 100 % hate those.*

As Hannah (Chapter 11 in this volume) stresses, proofs need to be evolved naturally. Hannah's belief is in line with Uhlig's (2002) idea of encouraging students to ask appropriate questions: "What happens if? Why does it happen? How do different cases occur? What is true here?" (p. 338). He believes if these questions are explored appropriately the knowledge can be gained in Theorems. A deep level of understanding can be achieved gently with the WWHWT sequence and can prepare students mentally and emotionally for the Definition-Lemma-Proof-Theorem-Proof-Corollary (DLPTPC) sequence. Uhlig believes that both instructor and student will not be satisfied by an early DLPTPC approach. As Thurston (1994, p. 163) stated: *"We are not trying to meet some abstract production quota of definitions, theorems and proofs. The measure of our success is whether what we do enables people to understand and think more clearly and effectively about mathematics."*

12.5 Conclusion

Research shows that for most students, the shift from school to linear algebra is not trivial. Although a typical first year linear algebra course contains a considerable amount of calculations, for example, finding the row echelon form of a matrix, calculating inverses, finding eigenvalues and eigenvectors, and finding the determinants, the course can get extremely sophisticated. Those students who have a tendency toward calculations (symbolic-matrix-action level) soon discover that learning the concepts well requires more than just computing matrices.

The ability to think in all three worlds of mathematical thinking and at the same time keeping up with mostly process level thinking is challenging. Furthermore, moving between the worlds at the right moment creates difficulties, specifically, some moves tend to be more challenging than the others (e.g., embodied to symbolic in eigenvectors, see Fig. 12.10).

The ability to solve more conceptual questions, dealing with nonroutine problems and proving theorems, does not come naturally to most students. These skills need to be fostered long before students arrive at college. As Harel and Sowder (2005) declare, advanced thinking in mathematics can potentially start as early as elementary school and must not wait until students take courses such as linear algebra. They believe, elementary and high school mathematics are rich with opportunities for students to develop advanced types of thinking.

References

Asiala, M., Brown, A., DeVries, D., Dubinsky, E., Mathews, D., & Thomas, K. (1996). A framework for research and curriculum development in undergraduate mathematics education. *Research in Collegiate Mathematics Education II, CBMS Issues in Mathematics Education, 6*, 1–32.

Briton, S., & Henderson, J. (2009). Linear algebra revisited: An attempt to understand students' conceptual difficulties. *International Journal of Mathematical Education in Science and Technology, 40*(7), 963–974.

Carlson, D. (1997). Teaching linear algebra: Must the fog always roll in? In D. Carlson, C. R. Johnson, D. C. Lay, A. D. Porter, A. Watkins, & W. Watkins (Eds.), *Resources for teaching linear algebra* (MAA notes, Vol. 42, pp. 39–51). Washington: Mathematical Association of America.

Day, J. M. (1997). Teaching linear algebra new ways. In D. Carlson, C. R. Johnson, D. C. Lay, A. D. Porter, A. Watkins, & W. Watkins (Eds.), *Resources for teaching linear algebra* (MAA notes, Vol. 42, pp. 71–82). Washington: Mathematical Association of America.

Dorier, J. L. (1990). *Continuous analysis of one year of science students' work, in linear algebra, in first year of French University*. Proceedings of the 14th Annual Conference for the Psychology of Mathematics Education, Oaxtepex, Mexico, II, pp. 35–42.

Dorier, J. L. (2000). *On the teaching of linear algebra*. Dordrecht: Kluwer Academic.

Dorier, J. L., Robert, A., Robinet, J., & Rogalski, M. (1994). *The teaching of linear algebra in first year of French Science University: Epistemological difficulties, use of the "Meta Lever" long-term organisation*. Proceedings of the 18th International Conference for the Psychology of Mathematics Education, Lisbon, Portugal, IV, pp. 137–144.

Dubinsky, E., & McDonald, M. (2001). APOS: A constructivist theory of learning. In D. Holton et al. (Eds.), *The teaching and learning of mathematics at university level: An ICMI study* (pp. 273–280). Dordrecht: Kluwer.

Duval, R. (2006). A cognitive analysis of problems of comprehension in a learning of mathematics. *Educational Studies in Mathematics, 61*, 103–131.

Hannah, J., Stewart, S., & Thomas, M. O. J. (2013). Conflicting goals and decision making: The deliberations of a new lecturer. In A. M. Lindmeier & A. Heinze (Eds.), *Proceedings of the 37th Conference of the International Group for the Psychology of Mathematics Education* (Vol. 2, pp. 425–432). Kiel, Germany: PME.

Hannah, J., Stewart, S., & Thomas, M. O. J. (2014). Teaching linear algebra in the embodied, symbolic and formal worlds of mathematical thinking: Is there a preferred order? In S. Oesterle, P. Liljedahl, C. Nicol, & D. Allan (Eds.), *Proceedings of the Joint Meeting of PME 38 and PME-NA 36* (Vol. 3, pp. 241–248). Vancouver, Canada: PME.

Hannah, J., Stewart, S., & Thomas, M. (2015). Linear algebra in the three worlds of mathematical thinking: The effect of permuting worlds on students' performance. In *Proceedings of the 18th Annual Conference on Research in Undergraduate Mathematics Education* (pp. 581–586). Pittsburgh, PA.

Hannah, J., Stewart, S., & Thomas, M. (2016). Developing conceptual understanding and definitional clarity in linear algebra through the three worlds of mathematical thinking. *Teaching Mathematics and its Applications: An International Journal of the IMA*. doi:10.1093/teamat/hrw001.

Harel, G., & Sowder, L. (2005). Advanced mathematical-thinking at any age: Its nature and its development. *Mathematical Thinking and Learning, 7*, 27–50.

Keith, S. (2001). *Visualizing linear algebra using Maple*. Prentice Hall.

Meel, D. E. (2005). Concept maps: A tool for assessing understanding? In G. M. Lioyd, M. Wilson, J. L. M. Wilkins, & S. L. Behm (Eds.), *Proceedings of the 27th Annual Meeting of the North American Chapter of the International Group for the Psychology of Mathematics Education*. Retrieved June 27, 2008, from http://www.allacademic.com/meta/p24713index.html

Sierpinska, A., Nnadozie, A., & Okta, A. (2002). *A study of relationships between theoretical thinking and high achievement in linear algebra*. Manuscript: Concordia University.

Stewart, S. (2008). *Understanding linear algebra concepts through the embodied symbolic and Formal worlds of mathematical thinking*. Doctoral thesis, University of Auckland. Retrieved from http://hdl.handle.net/2292/2912

Stewart, S., & Thomas, M. O. J. (2009). A framework for mathematical thinking: The case of linear algebra. *International Journal of Mathematical Education in Science and Technology, 40*(7), 951–961.

Stewart, S., & Thomas, M. O. J. (2010). Student learning of basis, span and linear independence in linear algebra. *International Journal of Mathematical Education in Science and Technology, 41*(2), 173–188.

Tall, D. O. (2008). The transition to formal thinking in mathematics. *Mathematics Education Research Journal, 20*, 5–24.

Tall, D. O. (2004). Building theories: The three worlds of mathematics. *For the Learning of Mathematics, 24*(1), 29–32.

Tall, D. O. (2010). Perceptions operations and proof in undergraduate mathematics. *Community for Undergraduate Learning in the Mathematical Sciences (CULMS) Newsletter, 2*, 21–28.

Tall, D. O. (2013). *How humans learn to think mathematically: Exploring the three worlds of mathematics*. Cambridge: Cambridge University Press.

Thomas, M. O. J., & Stewart, S. (2011). Eigenvalues and eigenvectors: Embodied, symbolic and formal thinking. *Mathematics Education Research Journal, 23*, 275–296.

Thurston, W. (1994). On proof and progress in mathematics. *Bulletin (New Series) of the American Mathematical Society, 30*(2), 161–177.

Uhlig, F. (2002). The role of proof in comprehending and teaching elementary linear algebra. *Educational Studies in Mathematics, 50*, 335–346.

Wawro, M., Sweeney, G., & Rabin, J. (2011). Subspace in linear algebra: Investigating students' concept images and interactions with the formal definition. *Educational Studies in Mathematics, 78*, 1–19. doi:10.1007/s10649-011-9307-4.

Wawro, M., Zandieh, M., Sweeney, G., Larson, C., & Rasmussen, C. (2011). *Using the emergent model heuristic to describe the evolution of student reasoning regarding span and linear independence*. Paper presented at the 14th Conference on Research in Undergraduate Mathematics Education, Portland, OR.

Index

S. Stewart (ed.), *And the Rest is Just Algebra*, DOI 10.1007/978-3-319-45053-7

MIX
Papier aus verantwortungsvollen Quellen
Paper from responsible sources
FSC® C105338

If you have any concerns about our products,
you can contact us on
ProductSafety@springernature.com

In case Publisher is established outside the EU,
the EU authorized representative is:
Springer Nature Customer Service Center GmbH
Europaplatz 3, 69115 Heidelberg, Germany

Printed by Libri Plureos GmbH
in Hamburg, Germany